Understanding American Agriculture

Challenges for the Agricultural Resource Management Survey

Panel to Review USDA's Agricultural Resource Management Survey

Committee on National Statistics

Division of Behavioral and Social Sciences and Education

NATIONAL RESEARCH COUNCIL
OF THE NATIONAL ACADEMIES

THE NATIONAL ACADEMIES PRESS
Washington, D.C.
www.nap.edu

THE NATIONAL ACADEMIES PRESS 500 Fifth Street, NW Washington, DC 20001

NOTICE: The project that is the subject of this report was approved by the Governing Board of the National Research Council, whose members are drawn from the councils of the National Academy of Sciences, the National Academy of Engineering, and the Institute of Medicine. The members of the committee responsible for the report were chosen for their special competences and with regard for appropriate balance.

This study was supported by Contract No. 53-3K06-05-1600 between the National Academy of Sciences and the United States Department of Agriculture. Support of the work of the Committee on National Statistics is provided by a consortium of federal agencies through a grant from the National Science Foundation (award number SBR-0453930). Any opinion, findings, conclusions, or recommendations expressed in this publication are those of the author(s) and do not necessarily reflect the views of the organizations or agencies that provided support for the project.

International Standard Book Number-13: 978-0-309-11092-1
International Standard Book Number-10: 0-309-11092-0

Additional copies of this report are available from the National Academies Press, 500 Fifth Street, N.W., Lockbox 285, Washington, DC 20055; (800) 624-6242 or (202) 334-3313 (in the Washington metropolitan area); Internet, http://www.nap.edu.

Printed in the United States of America.

Suggested citation: National Research Council. (2008). *Understanding American Agriculture: Challenges for the Agricultural Resource Management Survey*. Panel to Review USDA's Agricultural Resource Management Survey. Committee on National Statistics, Division of Behavioral and Social Sciences and Education. Washington, DC: The National Academies Press.

THE NATIONAL ACADEMIES

Advisers to the Nation on Science, Engineering, and Medicine

The **National Academy of Sciences** is a private, nonprofit, self-perpetuating society of distinguished scholars engaged in scientific and engineering research, dedicated to the furtherance of science and technology and to their use for the general welfare. Upon the authority of the charter granted to it by the Congress in 1863, the Academy has a mandate that requires it to advise the federal government on scientific and technical matters. Dr. Ralph J. Cicerone is president of the National Academy of Sciences.

The **National Academy of Engineering** was established in 1964, under the charter of the National Academy of Sciences, as a parallel organization of outstanding engineers. It is autonomous in its administration and in the selection of its members, sharing with the National Academy of Sciences the responsibility for advising the federal government. The National Academy of Engineering also sponsors engineering programs aimed at meeting national needs, encourages education and research, and recognizes the superior achievements of engineers. Dr. Charles M. Vest is president of the National Academy of Engineering.

The **Institute of Medicine** was established in 1970 by the National Academy of Sciences to secure the services of eminent members of appropriate professions in the examination of policy matters pertaining to the health of the public. The Institute acts under the responsibility given to the National Academy of Sciences by its congressional charter to be an adviser to the federal government and, upon its own initiative, to identify issues of medical care, research, and education. Dr. Harvey V. Fineberg is president of the Institute of Medicine.

The **National Research Council** was organized by the National Academy of Sciences in 1916 to associate the broad community of science and technology with the Academy's purposes of furthering knowledge and advising the federal government. Functioning in accordance with general policies determined by the Academy, the Council has become the principal operating agency of both the National Academy of Sciences and the National Academy of Engineering in providing services to the government, the public, and the scientific and engineering communities. The Council is administered jointly by both Academies and the Institute of Medicine. Dr. Ralph J. Cicerone and Dr. Charles M. Vest are chair and vice chair, respectively, of the National Research Council

www.national-academies.org

PANEL TO REVIEW USDA'S AGRICULTURAL RESOURCE MANAGEMENT SURVEY

BRUCE GARDNER *(Chair)*, Department of Agricultural and Resource Economics, University of Maryland
WALTER J. ARMBRUSTER, Farm Foundation, Oak Brook, Illinois
DAVID BINDER, Statistics Canada, Ottawa (Retired)
RAY D. BOLLMAN, Agriculture Division, Statistics Canada, Ottawa
CYNTHIA Z.F. CLARK, Methodology Directorate, Office for National Statistics, United Kingdom
FREDERICK CONRAD, Institute for Social Research, University of Michigan
ANI L. KATCHOVA, Department of Agricultural and Consumer Economics, University of Illinois at Urbana-Champaign
ARTHUR KENNICKELL, Microeconomic Surveys Unit, Board of Governors, Federal Reserve System
CATHERINE KLING, Department of Economics, Iowa State University
JEAN OPSOMER, Center for Survey Methodology, Iowa State University
BOBBY R. PHILLS, Small Fruits Program, Florida Agricultural and Mechanical University
STANLEY PRESSER, Sociology Department and Joint Program in Survey Methodology, University of Maryland
ROBERT D. TORTORA, Survey Research Center, The Gallup Organization, Washington, DC
MICHAEL K. WOHLGENANT, Department of Agricultural and Resource Economics, North Carolina State University

THOMAS J. PLEWES, *Study Director*
CARYN KUEBLER, *Associate Program Officer*
MICHAEL SIRI, *Senior Program Assistant*

Contents

Acknowledgments xi

Acronyms and Abbreviations xv

Executive Summary 1
Data Quality, 2
Conclusions, 2
Recommendations, 4
Investment in the Future, 5

1 Introduction 7
Survey Description, 8
Charge to the Panel, 10
Issues in Survey Operations, 11
The Panel's Approach, 12
Guide to the Report, 15

2 Contemporary Issues in American Agriculture 16
Priority Uses of ARMS Data, 17
Data Relevance, 25
Three Key Areas, 29
Maintaining Relevance by Developing New Data Sources, 45

3 Survey Management 49
Collaborative Management, 49
Managing the Changing Focus of ARMS, 55

4 Sample and Questionnaire Design 62
Survey Design Goals, 63
Sampling Frame, 65
Sample Design, 69
Questionnaire Design and Development, 74
Understanding Record-Keeping Practices, 79
Appropriateness of the Survey Questions, 81
Supplemental Data, 84

5 Data Collection 86
Collection Arrangements, 87
Quality Assurance, 88
Collection Methodology, 96
Data Capture, Editing, and Processing, 103
Metadata and Paradata, 104

6 Nonresponse, Imputation, and Estimation 107
Unit Nonresponse, 107
Item Nonresponse, 115
Imputation, 116
Estimation, 122

7 Methods for Analysis of Complex Surveys 127
Data Analysis Issues, 128
Evaluation of Current Approach, 131
Estimation Strategies for Complex Surveys, 132
Guide for Researchers, 146

8 Dissemination 150
New Data Sources and Products, 151
Microdata Access for Data Users, 154
Training for Data Users, 157
User Forums and Feedback, 159
Data Consistency, 159

9 Conclusions and Recommendations 162
Conclusions, 162
Recommendations, 164

References 170

Appendixes

A Panel Data-Gathering Activities 177

B Biographical Sketches of Panel Members and Staff 189

Acknowledgments

This report is the result of the work by many individuals, each of whom has contributed their expertise. This section acknowledges some of the central contributions to this extensive undertaking.

First, I acknowledge and thank my fellow panel members, who gave of their time and expertise so generously, participating in seven meetings and countless other activities critical to completing this report. They shared the administrative load in their service as session moderators during the open meetings and in their participation in the lively discussions of the issues at each of those events. Their assistance in drafting this report resulted in a more comprehensive and defensible product and is greatly appreciated.

Major activities of the panel were the four public sessions, three of which were organized and conducted as formal workshops. In developing and conducting these productive workshops, the panel enjoyed outstanding support from the staff of the National Agricultural Statistics Service (NASS) and the Economic Research Service (ERS). Robert Bass, director of the Census and Survey Division of NASS, and Mary Bohman, director of the Resource Economics Division, served as the primary agency liaisons and facilitated our work with admirable patience and perseverance.

Jim MacDonald of ERS provided much advice, counsel, and facilitation support and was instrumental in focusing the attention of the panel on the difficult issues related to tests of inference of the Agricultural Resource Management Survey (ARMS) data. He spoke before the panel on two occasions and maintained contact with the panel staff as the study emerged. Dania Ferguson of NASS assembled documents requested by the panel, some quite old and requiring considerable research, and prepared several responses to questions submitted by the panel.

In all, staff of NASS and ERS were of great assistance to us in understanding the complex survey design and implementation that characterizes ARMS. In addition to participating in the panel's open sessions, both NASS and ERS responded to many questions formally transmitted to the agencies. A summary of the panel's data-gathering activities appears in Appendix A.

The committee gratefully acknowledges the excellent work of the staff of the Committee on National Statistics (CNSTAT) and the National Research Council for support in developing and organizing the workshops and this report. The experience, wise counsel, and untiring efforts of Tom Plewes, study director for the panel, were indispensable in bringing this report to fruition. He was ably assisted by Caryn Kuebler of the CNSTAT staff, and Michael Siri provided administrative support. Michael Cohen assisted in clarifying and resolving issues related to analysis of complex survey data. William Greene provided sound advice on issues of inference in data analysis in service as an expert consultant to the panel.

We especially appreciate the participation of Constance F. Citro, director of CNSTAT, in the workshops and the preparation of this report. Her sage advice benefited the report in numerous ways.

This report has been reviewed in draft form by individuals chosen for their diverse perspectives and technical expertise, in accordance with procedures approved by the NRC's Report Review Committee. The purpose of this independent review is to provide candid and critical comments that assist the institution in making its report as sound as possible, and to ensure that the report meets institutional standards for objectivity, evidence, and responsiveness to the study charge. The review comments and draft manuscript remain confidential to protect the integrity of the deliberative process.

The panel wishes to thank the following individuals for their review of this report: Jason Abrevaya, Department of Economics, University of Texas at Austin; Richard N. Boisvert, Department of Agricultural Economics, Cornell University; D. Wade Brorsen, Department of Agricultural Economics, Oklahoma State University; Rachel Harter, National Opinion Research Center, Chicago, IL; Barrett Kirwan, Agricultural and Resource Economics, University of Maryland; Virginia Lesser, Survey Research Center, Department of Statistics, Oregon State University; Frank Potter, Statistical Services, Mathematica Policy Research, Inc., Princeton, NJ; and Alan M. Zaslavsky, Department of Health Care Policy, Harvard Medical School.

Although the reviewers listed above have provided many constructive comments and suggestions, they were not asked to endorse the conclusions or recommendations, nor did they see the final draft of the report before its release. The review of this report was overseen by Keith F. Rust, Westat,

Inc., Rockville, MD. Appointed by the NRC, he was responsible for making certain that the independent examination of this report was carried out in accordance with institutional procedures and that all review comments were carefully considered. Responsibility for the final content of the report rests entirely with the authoring committee and the institution.

Bruce Gardner, *Chair*
Panel to Review USDA's Agricultural
Resource Management Survey

Acronyms and Abbreviations

AAEA	American Agricultural Economics Association
ACS	American Community Survey
AEM	Association of Equipment Manufacturers
APHIS	Animal and Plant Health Inspection Service
ARDIS	Agricultural and Rural Development Information System
AREI	Agricultural Resources and Environmental Indicators
ARMS	Agricultural Resource Management Survey
BEA	Bureau of Economic Analysis
BLS	Bureau of Labor Statistics
CAI	Computer-Assisted Interviewing
CAPI	Computer-Assisted Personal Interviewing
CAR	Commodity Cost and Returns
CASIC	Computer-Assisted Survey Information Collection
CBO	Congressional Budget Office
CEA	Council of Economic Advisers
CEAP	Conservation Effects Assessment Project
CPS	Current Population Survey
CRP	Conservation Reserve Program
CRR	Costs and Returns Report
CSP	Conservation Security Program
CSV	Comma Separated Value
CV	Coefficient of Variation

DAG	Delete-a-Group
EDI	Electronic Data Interchange
EDR	Electronic Data Reporting
EPA	Environmental Protection Agency
EQIP	Environmental Quality Incentive Program
ERS	Economic Research Service, USDA
FAO	Food and Agriculture Organization of the United Nations
FAS	Foreign Agricultural Service, USDA
FO	U.S. Department of Agriculture State Field Office
FSA	Farm Service Agency, USDA
FTE	Full-Time Equivalent
FVS	Farm Value Sales
GDP	Gross Domestic Product
GEE	Generalized Estimating Equation
GHG	Greenhouse Gas
GIS	Geographic Information System
GPS	Global Positioning System
HEL	Highly Erodible Land
IDAS	Interactive Data Analysis System
IID	Independent and Identically Distributed
I-O	Input-Output
IRS	Internal Revenue Service
JPSM	Joint Program in Survey Methodology
MILC	Milk Income Loss Contract
NASDA	National Association of State Departments of Agriculture
NASS	National Agricultural Statistics Service, USDA
NIPA	National Income and Product Accounts
NOL	Non-Overlap
NRC	National Research Council
NRCS	National Resources Conservation Service, USDA
NRI	National Resource Inventory
OECD	Organisation for Economic Co-operation and Development
OEPNU	Office of Energy Policy and New Uses
OMB	Office of Management and Budget

P-B	Perry-Burt Procedure
PDA	Personal Digital Assistant
PPCR	Production Practices and Costs Report
PPI	Producer Price Index
PPR	Production Practices Report
PRISM	Project to Reengineer and Integrate Statistical Methods
PSM	Policy and Standards Memoranda
RDCs	Research Data Centers
RMA	Risk Management Agency, USDA
SAS	Statistical Analysis System Software
SCF	Survey of Consumer Finances
SIP	Sequential Interval Poisson
TCDO	Training and Career Development Office
TMDL	Total Maximum Daily Loads
USDA	U.S. Department of Agriculture

Executive Summary

The Agricultural Resource Management Survey (ARMS) is the federal government's primary source of information on the financial condition, production practices, and resource use on farms, as well as the economic well-being of America's farm households. ARMS data are important to the U.S. Department of Agriculture (USDA) and to congressional, administration, and industry decision makers when they must weigh alternative policies and programs that touch the farm sector or affect farm families.

ARMS was initiated in 1996 as a synthesis of existing USDA surveys on cropping practice, chemical use, and farm costs and returns. The survey is managed jointly by two USDA agencies: the Economic Research Service (ERS) and the National Agricultural Statistics Service (NASS). The three-phase annual survey is large, complex, and costly, with a budget of nearly $19 million in fiscal year 2006.

ARMS is unique in several respects. As a multiple-purpose survey with an agricultural focus, ARMS is the only representative national source of observations of farm-level production practices, the economics of the farm businesses operating the field (or dairy herd, greenhouse, nursery, poultry house, etc.), and the characteristics of the American farm household (age, education, occupation, farm and off-farm work, types of employment, family living expenses, etc.). No other data source is able to match the range and depth of ARMS in these areas.

American agriculture is changing, and the science of statistical measurement is changing as well. As with every major governmental data collection with such far-reaching and important uses, it is critical to periodically ensure that the survey is grounded in relevant concepts, applying the most up-to-date statistical methodology, and invested with the necessary design, estimation, and analytical techniques to ensure a quality product.

ARMS is a complex undertaking. From its start as a melding of data collected from the field, the farm, and the household in a multiphase, multiframe, and multiple mode survey design, it has increased in complexity over the decade of its existence as more sophisticated demands for its outputs have been made. Today, the survey faces difficult choices and challenges, including a need for a thorough review of its methods, practices, and procedures.

The Panel to Review USDA's Agricultural Resource Management Survey was established to conduct such a review, and this report is the product of its efforts. The panel's specific recommendations appear in context throughout the report and are presented together in Chapter 9.

DATA QUALITY

The panel focused on the elements of quality, broadly defined as "fitness for use," which for ARMS means relevance, accuracy, timeliness, accessibility, interpretability, and coherence. Assessing these elements of quality for ARMS required a review of its concepts, organization, sampling, questionnaire design, data collection, data processing, and dissemination. In doing so the panel has addressed issues of concern for ARMS and its uses for policy analysis and private-sector decision making in order to identify specific needed improvements, to outline testing and research to keep the survey current with data needs and state-of-the-art methods in the future, and in making it accessible to potential users.

The central qualitative dimension of survey data is accuracy. This is the element the panel was least able to assess with confidence, because knowledge is inadequate about the true values of many data items that ARMS estimates. Obtaining better information about the accuracy of survey-based information from the survey is a central reason for our call, sounded throughout this report, for a systematic program of methodology research and development, which would focus on questionnaire design, survey management, bias resulting from nonresponse, quality checks on responses, editing and imputation procedures, calibration to data from sources other than ARMS, and statistical procedures for calculating confidence intervals on estimates derived from ARMS data.

CONCLUSIONS

ARMS is an invaluable source of information on the current state of American agriculture, as well as the sole source of some important information on the linkages between fields, farms, and families that serves to illuminate the challenges faced by agriculture policy makers and farm families. Because the survey is so critical to understanding agriculture, it carries a special burden. Its methods, practices, and procedures must be designed to yield data of impeccable quality in view of their uses, and the data must be made available to the research community both inside and outside the federal government in order to generate the improved analytical knowledge the data makes possible.

At several points in this report, some of the methods and practices used in ARMS are characterized as "unique" or "unconventional." In large part, the unique nature of the survey is due to its complexity, with multiple modes and phases and with a goal to collect, classify, and aggregate several types of information from three interrelated but not entirely overlapping reporting units. ARMS also reflects some unique practices that are part of USDA's way of doing business, such as its board review process, which are not within the panel's purview to assess. Nonetheless, we have been able to document and assess the adequacy of the survey's design, data collection, analysis, and dissemination.

The panel concludes that ARMS has been carried out with admirable attention to achieving high standards on most of its elements. Nonetheless, there is room for improvement in all of them. In addition to identifying areas of needed improvement in current methods and practices, our review identifies several emerging challenges. These challenges are associated with the changing structure of farming, overall trends in federal surveys—such as the growing difficulty of obtaining satisfactory survey response—and the growing sophistication of survey data users, both inside and outside the federal government. NASS and ERS have attempted to respond to these challenges with some foresight—adding new questions, testing such initiatives as incentives for increasing reporting, developing proposals to collect longitudinal data, and enhancing the provision of microdata files in a protected environment. Our review leads to the conclusion that a number of areas need still further attention, and our recommendations can be considered a roadmap to the future for ARMS.

The panel also examined the appropriateness of the methods that ERS is using to fit statistical models to data from ARMS. The panel concludes that the current practice of NASS to provide survey weights with the ARMS data set, as well as the NASS and ERS recommendation to use the design-weighted approach for many of the analytical uses of the data, are appropriate and should be continued. However, the current one-size-fits-all approach for analytical inference for ARMS is somewhat restrictive, and users need more specific information on the sampling design and on nonresponse patterns and adjustments applied to the data sets. The panel concludes that a guide for researchers on how to approach certain commonly used analytical methods would improve the quality of analysis of the ARMS data.

RECOMMENDATIONS

Although appropriate attention is being paid to the basic elements of survey quality, much more can be done to improve important aspects of the survey. Issues that need more attention include the employment of analytical tools to investigate the quality of survey responses, additional control and further automation of the interview process, shift of focus from nonresponse rates to nonresponse bias, introduction of new methods of imputation of missing values and documentation of the results of imputation, improvements to variance estimation that will be more compatible with the types of data analysis that are now employed, and more attention to facilitating access to the data files for research and analysis.

The panel did not explicitly prioritize either the issues or our recommendations, although we do draw special attention to the need for an ongoing, joint, and appropriately funded methodology research and development program. We draw attention to special needs for a research and development program to support improvements in questionnaire design and development, unit and item nonresponse, and imputation and estimation. Such a program needs adequate resources both to support current and future research, development, and statistical analysis needs throughout the implementation of ARMS and to assess and manage the quality of the data. We also call attention to the need for better channels of communication with providers and users of the data. These initiatives will require an infusion of funding, and, in the case of ERS, enhancement of staff expertise in mathematical statistics and data analysis skills.

We are aware that our list of recommendations is long and that some of them will be costly to implement. Full implementation of all of them would require a significant fraction of the ARMS budget. In our view, if additional funds cannot be obtained, at least those recommendations involving research and development directly related to data quality assurance should be undertaken, even at the expense of reducing the size or scope of the survey. For other costly recommendations, notably the training programs and other services for data users outside USDA, additional funding could reasonably be sought, even from unconventional sources in the user community. For example, the land grant universities could be asked to support and perhaps to assist in implementing the training and data access improvements. The universities rely on other sources of USDA funding through the Cooperative Research, Education, and Extension Service, which might be interested in funding competitive National Research Initiatives or other grants for these purposes.

We recommend improvements to ARMS in all aspects of the survey:

- *Data integration and relevance,* including survey integration and longitudinal data.
- *Survey management,* including establishment of a research and development program, planning for changes in survey content, review by the NASS Advisory Committee on Agriculture Statistics, and stakeholder feedback and discussion.
- *Sample and questionnaire design,* including expanded use of generic clearance for survey testing from the Office of Management and Budget, cognitive or usability laboratory facilities, designation of subsamples for research and testing, consistency of variables, and formal collection of auxiliary information.
- *Data collection,* including paradata (i.e., data on the data collection process) and metadata (i.e., data on data items), the consequences of departures from standardized interview techniques, using available analytical tools to assess survey response quality, and moving to computer-assisted interviews and web-based data collection.
- *Nonresponse, imputation, and estimation,* including providing a foundation for appropriate response rate calculations, adjustment for response rates in the multiple phases, nonresponse bias, making the survey mandatory, changes to reduce item nonresponse, approaches to imputation, flagging missing data used for analysis, and clarifying the estimation process.
- *Methods of analysis,* including provision of sampling weights, using the design-weighted approach, jackknife replicates, design and nonresponse characteristics, in-house survey statistics expertise at ERS, and developing a guide for researchers.
- *Dissemination,* including improving the ARMS web tool, extending the availability of ARMS microdata, training for new data users, and database management practices.

INVESTMENT IN THE FUTURE

The ARMS program represents a significant investment of time, talent, respondent burden, and cost. It responds to congressional mandates to produce estimates of income for farms as business establishments by type of operation, to monitor the status of family farms, and to make annual estimates of the costs of producing commodities, such as wheat, feed grains, cotton, and dairy, covered under farm support legislation.

ARMS also enables USDA to meet its responsibility to support the U.S. national accounts by producing estimates of farm sector value-added and net income and the development of regional (state and county) accounts by provision of the state-level estimates of net farm income. It provides the data to implement USDA program requirements and to formulate and evaluate USDA polices and programs. The analytical program of ERS informs both public- and private-sector decision makers, through research and analysis on a wide variety of farm, farm household, environmental, and other rural issues, shedding light on important national issues like energy usage by supporting analysis of the impact of energy prices on farm production costs and, consequently, on the price of commodities.

As it begins a second decade of operation, improvements are needed in all aspects of the survey's operations to keep it vital. ARMS is too valuable to the nation for the U.S. Department of Agriculture not to make efforts now to increase its value for the future.

1

Introduction

The Agricultural Resource Management Survey (ARMS) program of the U.S. Department of Agriculture (USDA) is a study in complexity. Data are collected on three different occasions and from three different levels: the field, the farm, and the household associated with the farm. Other aspects, including the technical design (a multiphase, multiframe, stratified, probability-weighted sample), an ever-changing content, and the multiple modes of data collection increase the complexity.

Complexity is both a source of strength in the ARMS program and a challenge to those responsible for conducting the survey and analyzing the results. Today, the survey is more than an assemblage of several data collections within the National Agricultural Statistics Service. In the mid-1990s, ARMS was created by merging the objectives of two USDA surveys: the Farm Costs and Returns Survey (FCRS) and the Cropping Practices Survey (CPS). The FCRS objectives were to collect whole-farm production, organization and financial information, household demographic and financial information, and enterprise-level costs and returns information for selected commodities. The CPS objectives were to collect field-level chemical use, tillage practices, and other field practices for selected commodities. The ARMS program provides indispensable linkages of data for fields, farms, and households in a manner that permits analysis of management practices, profitability, and farm family composition and well-being, among other topics. No other source affords such a comprehensive and complete view of the American farm. Few other sources pose such complicated challenges to methodologists, data collectors, data processors, and analysts.

SURVEY DESCRIPTION

ARMS is USDA's primary source of information on the financial condition, production practices, and resource use of farms, as well as the economic well-being of America's farm households. Its data are essential to USDA and other federal administrative, congressional, and private-sector decision makers when they must weigh alternative policies and programs or business strategies that touch the farm sector or affect farm families (Box 1-1).

The basic USDA definition of a farm is any place from which $1,000 or more in nominal terms of agricultural products were produced and sold, or normally would or could have been sold, during the census year. This definition is common to both the Census of Agriculture and ARMS, and is reflected in the terms "census farms" and "census-defined farms." The definition has been steady for many years and encompasses many small, hard-to-measure businesses, which are difficult to identify and survey.

ARMS comprehensively provides observations of field-level farm practices, the economics of farm businesses operating the field (or dairy herd, greenhouse, nursery, poultry house, etc.), and the characteristics of the

BOX 1-1
A Thumbnail Sketch of the ARMS Program

The Agricultural Resource Management Survey is the primary source of information for the U.S. Department of Agriculture and the public on a broad range of issues about U.S. agricultural resource use, costs, and farm-sector financial conditions. It is the only source of information available for objective evaluation of many critical issues related to agriculture and the rural economy. The survey is conducted by the National Agricultural Statistics Service (NASS) in collaboration with the Economic Research Service (ERS).

The survey sample is designed to provide coverage of all farms in the 48 contiguous states plus state-level data for 15 major farm states. The population of farms, as defined by the Census of Agriculture, includes all establishments that produced and sold (or would normally or could have sold) at least $1,000 of agricultural products during the previous year. A sample from the NASS list frame is supplemented by a geographic sample of area tracts to ensure complete coverage.

ARMS collects data for whole farms and commodity-specific production practice and cost data, on a rotating basis, for selected commodities in Phase II of the survey and in commodity-specific versions of Phase III. The survey collects data in three phases:

- Phase I, the ARMS screening survey, collects general farm data, such as crops grown, livestock inventory, and the value of sales. Screening data are used to qualify (or screen) farms for the subsequent phases. Phase I is conducted from May through July.

American farm household (age, education, occupation, farm and off-farm work, types of employment, family living expenses, etc.) collected through interrelated, representative samples.

The survey has increased in complexity as it has matured over the years. The current pattern of rotating commodities between survey cycles is one example of a survey design decision that, although made for practical reasons, has tended to increase the complexity of the survey operation. Today, the pieces and parts of the survey can be described as

- a cooperative management and financing venture between two agencies that have independent objectives;
- a multiphase operation with three distinct survey operations, which, though integrated and building on each other, have different purposes and different constituencies;

BOX 1-1 continued

- Phase II collects data associated with agricultural production practices, resource use, and variable costs of production for specific commodities. Commodities are surveyed on a predetermined rotation, and up to five commodities are surveyed in a given year. Farm operators provide data on fertilizer and nutrient applications, pesticide applications, pest management practices, and irrigation. Phase II is conducted from September through December.
- Phase III collects whole-farm finance variables, operator characteristics, and farm household information. Farm operators provide data on farm operating expenditures, capital improvements, assets, and debt for agricultural production. In addition, operators provide data on farm-related income, government payments, the source and amount of off-farm income, and characteristics of themselves and their household. Phase III is conducted from February through April, with the reference period being the previous year. Respondents sampled in Phase II are asked to complete a Phase III report. Data from both phases provide the link between agricultural resource use and farm financial conditions.

Farm operators are selected to ensure adequate coverage by state and region and to minimize reporting burden. Strata are based on state, the value of agricultural sales, and type of farm. Phase I screening is performed by mail and phone. Operators who are in business or have the commodity of interest (which varies by year) are eligible to be selected for Phase II or Phase III. Phase II data are collected by means of personal interviews, while Phase III surveys are conducted using several modes of data collection (face-to-face, telephone, mailout-mailback with face-to-face follow-up for the mail respondents, and, on an experimental basis, the Internet).

The commodities surveyed are rotated every 5-6 years to focus on resource use and production costs for specific commodities.

SOURCE: National Agricultural Statistics Service.

- an elaborate sampling frame consisting of both a traditionally constructed list of farms (the list frame) and a special list obtained by an intensive geographic area screening (the area frame);
- a complex collection scheme implemented under a cooperative agreement by a nonfederal organization;
- an ambitious, cognitively challenging survey questionnaire that, despite several efforts at simplification, is perceived by USDA to be so burden some to respondents that pains are currently taken to minimize revisits to them, which limits the ability to longitudinally follow these reporting units; and
- a complex estimation and variance computation procedure, which, although appropriate for its purpose, can place limitations on the ability of data analysts to perform multivariate analysis using standard statistical packages and to determine if the analytical result is statistically valid and reliable.

CHARGE TO THE PANEL

The responsible agencies, the National Agricultural Statistics Service (NASS) and the Economic Research Service (ERS) of the U.S. Department of Agriculture, are well aware of the challenges posed by the ARMS program and have sought to improve many aspects of survey operations and analysis as time and resources have permitted. As part of a program of continuous improvement, the two agencies joined in requesting this review of ARMS. To conduct the review, National Research Council, through the Committee on National Statistics, appointed the Panel to Review USDA's Agricultural Resource Management Survey, whose members have expertise in household and business survey methods, the economics of farming and farm households, and complex sample designs.

The charge to the panel was to address two related tasks: (1) review the characteristics of the survey itself, including concepts, sample design, questionnaire design, data collection, and data processing and estimation, considering for each whether USDA is using the best practices in each of these aspects of the survey and how its practices might be improved, and (2) study the uses of the data for econometric policy-relevant analyses. Of particular concern is whether ARMS uses state-of-the-art methods to fit statistical models to ARMS data—that is, for estimating the variance of estimates and the implications for univariate and multivariate estimation of the complex sample design. Drawing on its experience as major users of ARMS data, the panel also reviewed the processes of the survey and various means of expanding access to the microdata for econometric and other analysis.

ISSUES IN SURVEY OPERATIONS

The complex nature of ARMS raises issues in nearly every aspect of survey operations, from conceptualization, organization, sampling, and questionnaire design, to data collection, data processing, analysis, and dissemination.

NASS and ERS have raised a number of questions about:

1. the adequacy and utility of the area frame sample, which supplements a list frame sample from the Census of Agriculture and other sources;
2. the appropriateness of the process used to determine questionnaire content, which includes operating characteristics and business financial information for each farm, the farm operator's household off-farm income and other characteristics, and operating characteristics and farming practices for specific field crops, surveyed in various years;
3. whether best practices are being used to elicit high-quality responses to the economic and demographic questions, whether the ARMS questions pose major challenges to high-quality responses in that they are sensitive from a privacy perspective and very detailed in their inquiries about resource allocations and economic outcomes for the farm business and the farm household, and the effects of the fact that questions require extended memory recall and family records;
4. whether best practices are being used to elicit economic measures of farm and household performance for the prior year or, in some cases, the previous year;
5. how trade-offs between respondent burden and the need for imputation should be evaluated, and whether the best methods for imputing data due to nonresponse or due to unasked questions for particular subsamples are being employed;
6. other possible approaches for further reducing individual item nonresponse for both the mail and the in-person versions of the survey; and
7. whether respondents' comprehension of, and responses to, ARMS questions are consistent with the concepts that USDA intended to measure and, if not, what additional information could be informative.

The panel's assessment of the current methodology and practices has been conducted in light of an understanding of the state of survey methodology and best practices. These issues are addressed in Chapters 4, 5, and 6.

In addition to publication of summary data in cross-sectional and time series form, use of the data is expanding for hypothesis testing, econometric modeling, and other methods contributing to policy analysis. This growing role is the result of an ongoing, successful program to promote awareness of, and access to, ARMS data by ERS. Today, a substantial and expanding number of government and academic researchers are using the survey data to conduct research on a wide range of topics, including analysis of farm business and household responses to government programs.

As the uses of ARMS data for increasingly sophisticated purposes have expanded, the experts in data analysis at ERS have increasingly been called on to provide advice and guidance to internal and external data users on appropriate techniques and methodologies for data analysis. To assist ERS, the panel has reviewed statistical hypothesis-testing procedures using ARMS data. We consider USDA's choice of statistical procedures for estimating standard errors to test hypotheses with simple estimates and with complex econometric models and make recommendations on best practices for variance estimation and other statistical issues in the use of ARMS data by policy analysts. The types of specific questions that are considered include, for univariate statistics: the appropriate methods for calculating standard errors for use in hypothesis testing; the adequacy of the delete-a-group jackknife variance estimator for calculating standard errors, in general and in small samples; possible improvements to the delete-a-group jackknife estimator; and effects of ignoring the survey design in hypothesis testing. For advanced multivariate methods, the issues are more complex: the consequences of ignoring survey design when one is using the data in econometric analyses; the need to account for survey design in estimation of coefficients and standard errors and, if so, what approaches to take; the impact of the delete-a-group jackknife variance estimator on hypothesis testing for policy inferences and for professional presentation; any weaknesses of the jackknife estimator in this context; and whether alternative complex-sample estimators, particularly those used in standard analytic programs such as STATA or SAS, are acceptable. These issues are considered in Chapter 7.

THE PANEL'S APPROACH

Appointed in January 2006, the panel held its first meeting on February 2-3, 2006. Over the course of its inquiry, the panel has conducted five open meetings that involved interaction with ERS and NASS staff, as well as key data users, policy makers, and additional technical experts: a workshop on statistical methodology, June 8-9, 2006; a session for data users at the annual meeting of the American Agricultural Economics Association, July 24, 2006; a workshop on concepts and measurement, September 28-29, 2006; a workshop on inference, December 7-8, 2006; and an open discussion of cost-of-production issues, January 18, 2007. The agendas of these meetings appear in Appendix A.

The panel's recommendations respond to USDA's concerns about ARMS and its uses for policy analysis, identify specific needed improvements, and suggest a program of research, testing, and development to keep the ARMS current with data needs and state-of-the-art methods in the future. With the time and resources available, it was not possible to formulate a particular solution to each of the issues, some of which require considerable research, development, and testing. However, we did identify several issues for priority review in the research and development program we recommend. These research recommendations appear throughout the report.

The focus in this report is on quality, broadly defined as "fitness for use." The definition of quality throughout the international statistical system consists of several constituent elements or dimensions. One commonly used set of six elements, to which this report adheres, includes relevance, accuracy, timeliness, accessibility, interpretability, and coherence (Organisation for Economic Co-operation and Development, 2003; Statistics Canada, 2003).

The relevance of statistical information reflects the degree to which it improves the decisions of clients. It is concerned with whether the available information improves the value of a decision and sheds light on the issues that are important to users. Assessing relevance is subjective and depends on the varying needs of users. The statistical agency's challenge is to weigh and balance the conflicting needs of current and potential uses to produce a program that goes as far as possible in satisfying the most important needs within given resource constraints. Relevance is considered in Chapter 2 in the context of the need to make decisions on contemporary issues in American agriculture.

The accuracy of statistical information is the degree to which the information correctly describes the phenomena it was designed to measure. It is usually characterized in terms of error in statistical estimates and is traditionally decomposed into bias (systematic error) and variance (random error) components. It may also be described in terms of the major sources of error that potentially cause inaccuracy (e.g., coverage, sampling, nonresponse, response). The major potential sources of error are discussed in Chapters 4, 5, and 6.

The timeliness of statistical information refers to the delay between the reference point to which the information pertains and the date on which the information becomes available for use. There is typically a trade-off between timeliness and accuracy. Timeliness, in terms of its influence on relevance, is addressed in Chapter 2, while Chapter 8 discusses timeliness as a factor in overall quality.

The accessibility of statistical information refers to the ease with which it can be obtained from the statistical agency. This involves issues of dissemination, which are covered in Chapter 8, including the ease with which the existence of information can be ascertained, the suitability of the form in which the information can be accessed, and the availability of user support services. The cost of using the information, including both direct costs for data products and the cost of travel to centralized repositories of research data, may also be an aspect of accessibility for some users.

The interpretability of statistical information reflects the availability of supplementary information about the data, often in the form of metadata (i.e., data about data items) and paradata (i.e., data about the data collection process), which are necessary to interpret and use it appropriately. This information usually includes the underlying concepts, variables and classifications, the methodology of data collection and processing, and indications or measures of the accuracy of the statistical information. Chapter 5 discusses these components of quality.

Finally, the coherence of statistical information reflects the degree to which it can be successfully brought together with other statistical information within a broad analytic framework and over time. The use of standard concepts, classifications, and target populations promotes coherence, as does the use of common methodology across surveys. ARMS presents special challenges for coherence, in that there is a need to be internally coherent among the three phases of the survey, as well with other USDA data. We examine one of the methods for achieving coherence—the calibration process used for setting production estimates and aligning the various sources of data regarding production in Chapter 6, as well as the methods for coherence in analysis of the data from ARMS itself, in Chapter 7.

At the operational level, these elements of quality are reflected in guidelines that have been promulgated by the U.S. Office of Management and Budget (OMB) (U.S. Office of Management and Budget, 2001). In addition to quality, the OMB guidelines address utility, objectivity, integrity, transparency, and reproducibility of information disseminated by federal agencies.

These guidelines have been appropriately embraced by the leadership of both NASS and ERS. This is important, since quality assurance is mainly a management function. Along with leaders of several of the largest federal statistical agencies, the administrators of NASS and ERS signed a statement in 2002 delineating federal statistical organizations' guidelines for ensuring and maximizing the quality, utility, objectivity, and integrity of disseminated information. The role of the statistical agency in ensuring quality is summarized in this statement and bears repeating as the underlying theme of this report (U.S. Office of Management and Budget, 2002, p. 38468):

> A statistical organization's commitment to quality and professional standards of practice further includes: the use of modern statistical theory and practice in all technical work; the development of strong staff expertise

> in the disciplines relevant to its mission; the implementation of ongoing quality assurance programs to improve data validity and reliability and to improve the processes of compiling, editing, and analyzing data; and the development of a strong and continuing relationship with appropriate professional organizations in the fields of statistics and relevant subject-matter areas.
>
> To carry out its mission, a Federal statistical organization assumes responsibility for determining sources of data, measurement methods, methods of data collection and processing while minimizing respondent burden; employing appropriate methods of analysis; and ensuring the public availability of the data and documentation of the methods used to obtain the data. Within the constraints of resource availability, a statistical organization continually works to improve its data systems to provide information necessary for the formulation of public policy.

Beyond this, the OMB has directed each federal agency to issue its own information quality guidelines, and further guidelines have been issued by USDA, NASS, and ERS (U.S. Department of Agriculture, 2006; National Agricultural Statistics Service, 2007a; Economic Research Service, 2003).

More recently, OMB has issued detailed *Standards and Guidelines for Statistical Surveys* (U.S. Office of Management and Budget, 2006b), a comprehensive guide to developing and managing surveys in such a way as to obtain OMB approval for their conduct. In these guidelines, quality standards for the various stages of survey operations have been spelled out in some detail. The topics covered range from satisfactory survey response rates to the development of sampling frames to drawing of inferences from the data. The panel refers to these standards and guidelines in the report when discussing issues of compliance and noncompliance.

GUIDE TO THE REPORT

Following this introduction, we lay out the contemporary issues facing American agriculture and the relevant uses of ARMS data to address them in Chapter 2. These uses include those driven by congressional mandates and by agency and research community needs. Chapter 3 outlines the organizational structure behind ARMS and the collaborative management of NASS and ERS. Issues of sample and survey design, data collection, nonresponse, imputation, and estimation are addressed in the next three chapters. Chapter 7 provides a framework for analysis of complex surveys and issues related to inclusion of survey design in estimation. Data user concerns, including dissemination of data and opportunities for user feedback and training are addressed in Chapter 8. Chapter 9 summarizes the panel's conclusions and recommendations.

2

Contemporary Issues in American Agriculture

The Agricultural Resource Management Survey (ARMS) has been called the mirror in which American farming views itself (Economic Research Service, 2007). In its detail, the survey provides myriad observations about farms and farm households. When assembled into aggregate measures, it aims to form a comprehensive picture. The goal of providing in-depth detail that adds up to a multidimensional view of agriculture is a unique and important aspect of the survey.

The value of ARMS is in its totality and the interactive nature of the data, not just the value of each data item. It is a comprehensive documentation of many aspects of U.S. agriculture and farm households. In part because it is so comprehensive—and unique in that respect—it has important uses in illuminating contemporary issues in American agriculture. These uses form the basis for assigning priorities of effort, justifying the resources that are devoted to the survey, and largely defining its content and products.

This chapter begins with an illustrative catalogue of mandated, programmatic, and research uses of ARMS data by the U.S. Department of Agriculture (USDA) and other federal agencies, the private sector, academic researchers, and the general user community. It continues with a general discussion of contemporary issues in American agriculture and how they generate a demand for information that will help policy makers and private-sector decision makers at all levels understand and deal with those issues. It goes on to discuss in more detail three specific areas of interest in agriculture policy: environmental and resource management, commodities, and the economic situation of farms and farming families. The discussion includes an assessment of the adequacy of the ARMS data to illuminate these issues.

PRIORITY USES OF ARMS DATA

The ARMS program represents a significant investment of time, talent, respondent burden, and resources. To justify this investment, the survey must be responsive to a set of core requirements that address legislative, programmatic, and analytical needs. These core requirements build on those of the predecessor surveys, which conveyed into ARMS when it was established in 1996, and have been supplemented by more contemporary and changing requirements.[1]

The task of meeting these core requirements translates into a series of priorities for the ARMS program. The data items needed to meet the core requirements have largely been maintained and protected by making sure these items are included before any other items are added. For the National Agricultural Statistics Service (NASS) and the Economic Research Service (ERS), these priorities affect the content of the questionnaires, which in turn are instrumental to the survey's ability to meet the core requirements.

Mandated Uses

USDA is required by Congress, through both authorizing and appropriation legislation, to produce a sizeable portion of the data items that are included in ARMS. Cost-of-production data are required by several pieces of legislation, and one piece of legislation is very specific. The U.S. Code states that the "Secretary of Agriculture, in cooperation with the land grant colleges, commodity organizations, general farm organizations, and individual farmers, shall conduct a cost of production study of the wheat, feed grain, cotton, and dairy commodities under the various production practices and establish a current national weighted average cost of production. This study shall be updated annually and shall include all typical variable costs, including interest costs, a return on fixed costs, and a return for management" (U.S. Code, Title 7).

Environmental and food safety legislation call for data on chemical use on field crops. The Food, Agriculture, Conservation, and Trade Act of 1990 and the Food Quality Protection Act of 1996 require NASS to collect data on field crop chemical use and publish those data annually (in the Agricul-

[1]The predecessor surveys of ARMS were the Farm Costs and Returns Survey (FCRS) and the Cropping Practices Survey. The predecessors to the FCRS were the Farm Production Expenditures Survey and the Cost of Production surveys. The household component was added to the FCRS in 1988 with the Farm Operator Resource version.

tural Chemical Usage Field Crops Summary and the Agricultural Chemical Usage Restricted Use Pesticide Summary).

Some data series are used in the preparation of mandated reports. For example, in preparing the Annual Report on Family Farms required by the Food and Agriculture Act of 1977, ERS draws on ARMS data for information on a host of relationships, including (1) farm participation in agricultural programs and the distribution of farm program payments; (2) the structure and organization of farms, including family and nonfamily ownership; (3) the use of new production technologies and other management practices; (4) farm use of credit; (5) farmers' participation in off-farm employment; and (6) identifying the characteristics of producers purchasing crop insurance.

ARMS data are sometimes input to other data series, thus fulfilling their mandate as a derived requirement from another product. Mandated uses of ARMS data include annual estimates of average income for U.S. farm operator households, and annual cost-of-production estimates for over 15 agricultural commodities. These mandated data items generate further requirements, in that a number of data items are necessary to compute these derived estimates.

Similarly, ARMS production input data provide annual weights for the NASS computation of the Prices Paid by Farmers Index. In turn, this index is used to calculate parity prices required by the 1933 Agricultural Adjustment Act. Parity prices help administer federal marketing orders for some 45 fruits, vegetables, and nuts. The indices are also required by the 1978 Public Range Improvement Act to calculate annual federal grazing fees on the nation's western public lands by the U.S. Bureau of Land Management and the U.S. Forest Service.

National Accounts

USDA is required to support the U.S. national accounts by producing estimates of farm-sector value-added and net income. The agency responsible for the national accounts, the Bureau of Economic Analysis (BEA), reports that the primary source for the input-output (I-O) estimates of farm output is "cash receipts from farm marketings," which is compiled from ARMS and various sources by ERS. This series is considered to be of higher quality and of more relevance for the I-O estimates than the data collected in the Census of Agriculture. The I-O estimates of farm expenses are based on the ARMS farm costs and returns survey (Horowitz and Planting, 2006). BEA uses USDA's annual estimates of net farm income to develop its annual estimates of gross domestic product and personal income. ARMS data also contribute to the development of BEA's regional (state and county) accounts through inclusion of the state-level estimates of net farm income. The Farm

Production Expenditures by type of organization for county income distribution is produced through a standing request each year from BEA for publication in *County Business Patterns*.

USDA Programmatic Needs

ARMS data provide a basis for updating weights in the Prices Paid Index Program,[2] as well as for data to implement USDA program requirements and to formulate and evaluate USDA polices and programs. For example, publication of income for farm households was required by a past secretary of agriculture. The departmental definition of a limited resource farm relies on ARMS data. Fuel expenditures data are also used by the USDA Office of Energy Policy and New Uses to assess agricultural fuel usage.

General Information

ARMS provides data to inform decision makers on a wide variety of farm, farm household, environmental, and other rural issues. For example, ARMS data shed light on important national issues, such as energy use, by supporting analysis of the impact of energy prices on farm production costs and, consequently, on the price of commodities. ARMS data are also used to (1) gather information about the relationships among agricultural production, resources, and the environment; (2) determine the costs to produce various crop and livestock commodities and the relative importance of various production expense items; (3) help determine the net farm income of farmers and ranchers and provide data on the financial situation of farm and ranch businesses, including debt levels; and (4) help determine the characteristics and financial situations of farm and ranch operators and their households, including information on management strategies and off-farm income.

Research and Analysis

ARMS data support the ERS program of research and analysis, which depends on these data, and contribute to other research and analysis work by providing basic cost-of-production and input use information. The data also support an active and growing program of research and analysis at academic institutions and other organizations that aim to illuminate diverse issues in contemporary American agriculture. Many of these studies are conducted as collaborative efforts between ERS and the research community.

[2]Prices paid indices are used to compute parity prices under the Agricultural Adjustment Act of 1938 as amended, Title III, Subtitle A, Sections 301a. The Agricultural Marketing Service uses prices paid indices to determine support prices.

ERS is both a producer and a consumer of ARMS data. In addition to its extensive inventory of data sets and recurring publications, the ERS generates demands for ARMS data in its selection of topics for analytical projects. These demands often form the basis for additions and modifications in the data collected in ARMS. To illustrate, ERS work plans for 2006 included several initiatives on topics as varied as farm structure and governance, to food safety, to program participation, to farm household behavior, each using the ARMS as a major data source:

- *Changing structure of livestock and poultry industries.* Successive questionnaire versions targeted hogs, dairy, and broiler (chicken raised for meat) producers in 2005-2007 about their operations in the prior year. ERS intends to use the data to summarize structural change in each industry; analyze the sources of production cost differences, such as location or scale economies, as drivers of structural change; derive baseline information on manure management practices; and assess the effects of structural change and manure management practices on excess nutrient loadings.
- *Food safety strategies in the livestock and poultry supply chain.* ERS has included questions on production practices related to food safety and their links to processor/contractor requirements in the survey as noted above. ERS intends to use the data to summarize the use of various practices and to identify their impacts on production costs.
- *Producer responses to changes in federal peanut and tobacco programs.* The 2004-2006 Cost and Returns Report versions and the 2004 peanut version of ARMS included questions designed to elicit information on how producers responded to the fundamental changes in federal support for those industries, to be used in broader ERS analyses of the market response to the changes.
- *Farm operators' input procurement strategies.* Farmers may procure inputs through purchase, lease/rent, service contracting, on-farm provision, or joint sharing of inputs with other producers. Questions in the Costs and Returns Report allow ERS to identify those several strategies, to generate baseline estimates of their use, and to assess the effects of each strategy on farm efficiency and profits.
- *Farm governance and organization.* Large farms may have complex organizational structures; those structures may be an effective means of organizing farm production, but they may also provide a means to avoid USDA program rules. They also complicate the assessment of the role of family farms in American agriculture. Starting with the 2005 Costs and Returns Report, ERS has added questions designed to provide a more precise picture of organizational structures in American agriculture.
- *Farm and farm household linkages to rural communities.* The 2004 Costs and Returns Report contained questions pertaining to the linkages

between farms and farm households and nearby communities to better understand the importance of farm policy to rural communities and the importance of general economic policy to farm households.

- *Health and health care of rural residents and farmers.* The 2005 Costs and Returns Report contained additional questions on the health insurance coverage of each member in the farm operator household, the sources of the insurance (employer, Medicare, private purchase by the household, etc.), and household health insurance costs and out-of-pocket health expenditures. ERS will complement the ARMS information for farm households with information from the Current Population Survey, the National Health Interview Survey, and other sources to characterize outcomes for other rural households.
- *Consumption behavior of farm households.* The 2004 Costs and Returns Report reinstated a revised version of questions on household living expenses, and the 2005 report contained a further revised version to calculate consumption flows for the household. The goal was to achieve consistency with the measures of "consumption expenditures" and "current (annual) consumption service flows for the household" calculated with the Consumer Expenditure Survey of the Bureau of Labor Statistics, collected for a nationally representative sample of U.S. households.
- *Design options for green payments programs.* Proposed "green payments" would alter federal farm policy to jointly achieve income and conservation goals with the same payment instrument. ERS will compare various specific alternatives, using a "screening model" that links ARMS Phase III data to environmental and conservation program data. The screening model will compare alternative green payments to estimates of farmers' willingness to accept payments for various conservation practices.

ARMS is the only source of nationwide data to support research on farmers' decisions to adopt new technologies and to relate those decisions to (1) the economic performance and structural attributes of farms and farm families, and (2) the subsequent environmental impacts. Key technology adoption decisions being tracked include the choice of bioengineered seed, the selection of waste management practices by livestock producers, the use of chemical and biological pest management alternatives, and the use of information management technologies ranging from precision farming in crop production to marketing commodities and buying inputs via the Internet.

Federal Government Agencies

Representatives from several federal agencies briefed the panel on the value and uses of ARMS. Common uses in these agencies include analysis of conservation practices, payments, ethanol trade, distribution of commodity subsidies and alternatives, pesticide uses and regulations, and estimates of income and cost of production. Many noted the tie to farm households as a unique benefit of the survey. For the most part, these agencies do not make direct use of the microdata. They have asked for special reports from ERS, used the public-use aggregates available in the online ARMS briefing room, and used information derived from published reports and briefing papers that used the microdata.

The *Congressional Budget Office* (CBO) provides nonpartisan budget information for the agriculture committees in Congress through a small agricultural team (three employees in 2006) in the budget analysis division. This team produces mandated cost estimates for every bill coming out of the full committees, and it does a number of estimates before the bill gets to that point. CBO provides baseline estimates twice a year, including the current year plus 10 years, and tracks changes over time. The team needs data mainly on incentives to adopt and levels of adoption of conservation practices and general details on payments. It obtains its information indirectly through outside reports or conversations with people who know the data. CBO staff, who do not have access to the ARMS microdata, find briefing papers useful due to tight time constraints and stress the importance of ARMS for information on the means testing and diversity of farm operations.[3]

The President's *Council of Economic Advisers* (CEA) has used the ARMS aggregate data directly from the website as well as other information from USDA to assess both short-term and long-term changes in agriculture. Although it is difficult to anticipate its needs, some examples of prior topics include trade of ethanol and distribution of commodity subsidies and alternatives. ARMS cost-of-production data can assist with the latter of these two.

In a presentation to the panel, an agriculture specialist formerly on the CEA staff expressed an interest in ARMS allowing stratification of data by government payment, commodity payment (receipt/nonreceipt), and subsidy payment class. He suggested that more value might be gained from the standpoint of commodity programs by decreasing the sampling rate of smaller farms. About half of the farms in the Census of Agriculture have less than $10,000 of gross revenue, and about 9 percent of census farms account for over 90 percent of production. ARMS collects financial and income information in the same questionnaire, allowing analysts to cross-tabulate assets, debts, and net income. This cannot be done with other data sources, including the Census of Agriculture.[4]

The *Environmental Protection Agency* (EPA) is concerned with pesticide practices regarding fruit and vegetable crops, which are commodities

[3]Presentation by James Langley, Congressional Budget Office, September 28, 2006.
[4]Presentation by Joe Cooper, ERS and past CEA staff member, September 28, 2006

that ARMS does not cover. Typically the agency needs to determine the impact on agriculture and the total economy if a pesticide is registered or cancelled or an allowable use is changed in any way. In order to accomplish this, information is needed on the use of the pesticide under consideration, the use of pesticides that can be considered alternatives to that pesticide, and the yield and quality impacts of changes in pesticide use.

EPA gets most of these data from other NASS sources, as well as Doane Marketing Research, a proprietary data source with 17,000 respondents, primarily from larger farms, that provides estimates comparable in many respects to NASS estimates, although not as complete as NASS on the timing of pesticide application.

Data on how much pesticide a farm uses, on what crops it is used, and the timing of pesticide applications and the target pests of those applications are important to EPA. If these basic information items were available from ARMS, they would provide a key framework for many analyses. EPA currently makes extensive use of crop budgets, but they are limited in that they are not able to be cross-tabulated with information about actual conditions—the quantity and condition of crops or the timing of pesticide applications.[5]

The staff in the USDA *Office of the Chief Economist* primarily works on short-run, day-to-day issues, drawing on resources throughout the department. They frequently make requests to ERS for ARMS data and analysis to produce farm income and cost-of-production estimates, for example:

- Calculating farm energy cost-to-output ratios;[6]
- Highlighting the distribution of farm income, household income, and potential problems servicing debt;
- Explaining the distribution of farm program payments; and
- Identifying the characteristics of producers purchasing crop insurance.

An important benefit of ARMS data is their tie to the households of farm operators. The financial data of ARMS assists the Office of the Chief Economist in evaluating the economic impact of policies, and there is a lot

[5]Presentation by Arthur Grube, Environmental Protection Agency, September 28, 2006.

[6]USDA chief economist Keith Collins testified to the Senate Agriculture Committee on the implications for U.S. agriculture of higher energy prices using ARMS data to show how energy price increases would affect production costs for different commodities and regions. The testimony provided an estimate to the committee, also relying on ARMS data, of the likely impacts of energy price increases and hurricane damage on farm incomes. Finally, ARMS data were used to suggest likely substitution responses in farm inputs. Statement of Keith Collins, Chief Economist, U.S. Department of Agriculture before the U.S. Senate Committee on Agriculture, Nutrition, and Forestry, November 9, 2005, <http://www.usda.gov/oce/newsroom/congressional_testimony/collins_11092005.doc>.

of interest in the off-farm income of household members. In a presentation to the panel, the office representative stated that it would use the data more extensively if they were available longitudinally; he also noted the value of distributional information from ARMS and data on the new technology adoption and how it affects the cost of production.[7]

The *National Resources Conservation Service* (NRCS), previously the Soil Conservation Service, is a USDA agency that provides leadership in a partnership effort to help conserve, maintain, and improve the country's natural resources and environment. It analyzes the interplay between rural land and land under development.

NRCS has used ARMS data on production costs, input use, and technology adoption in selected regions to assess the performance of conservation programs, and it has used information on manure and fertilizer use and management practices to calculate nutrient mass balances for selected crops, regions, and management systems. The agency has consulted ARMS for hog production characteristics, practices, and costs to assess the economic feasibility of proposed nutrient management policy alternatives. It used ARMS data about incentives for farmers' use of resource management practices to assess the effect of their programs.

Use of ARMS data by *other USDA agencies* is both periodic and topical. USDA reports the following examples of uses by other USDA agencies:

- The Agricultural Marketing Service uses ARMS data in deriving its monthly cost-of-production estimates for milk production for the United States and five regions.
- The Risk Management Agency has used special tabulations from ARMS to understand levels of farm income and risk management tools used by farmers.
- The Agricultural Research Service has used special tabulations from ARMS to better understand the structural and production characteristics of farms and the demographic characteristics of farm operators for each of its research planning regions.
- The Farm Service Agency uses the annual burley and flue-cured tobacco cost-of-production estimates derived from ARMS data to help set tobacco price support levels.
- The Cooperative State Research Education and Extension Service has used ARMS information about adoption of alternative pest management strategies by farmers in its program planning.
- The Rural Business-Cooperative Service has used ARMS data to obtain information about the use of cooperatives by farmers.

[7]Presentation by Joseph Glauber, U.S. Department of Agriculture, September 28, 2006.

Interest Groups

There is a wide variety of interest group users with a wide variety of data needs. Two of the groups, the American Farm Bureau Federation and the National Cotton Council, participated in a panel discussion of data needs and the adequacy of the ARMS program for meeting those needs.

The *American Farm Bureau Federation (AFBF)*, an independent, nongovernmental organization governed by and representing farm and ranch families, considers ARMS one of the most important USDA products. It sends news releases to about 75,000 members and strongly encourages its members to respond to NASS surveys. With the reauthorization of Farm Bill legislation scheduled for 2007, ARMS may become more vital in terms of measuring costs of production and net farm revenues. The AFBF expressed an interest in more information on contracting and information on individual commodities on a more frequent basis.[8]

The *National Cotton Council* is typical of many commodity-oriented interest groups. It uses the ARMS on a regular basis, as it is the best source for many of its data needs. The cost-of-production information is used most often and gets a lot of attention in the industry. With a 1997-2003 gap in the ARMS rotating collection of cotton cost-of-production data, there are doubts as to whether the adoption of new practices (e.g., biotechnology) is appropriately captured as estimates are updated in the intervening years. Besides wanting more frequent collection of cotton cost-of-production data in the ARMS rotation, the council would like to see the ARMS sample size increased to enable the provision of data on as local a geographic level as possible, in order to have distributional information beyond averages.

The National Cotton Council does not conduct any elaborate in-house surveys, although they do conduct a survey on acreage intentions at the beginning of each year. Like the AFBF, they encourage members to respond to all NASS surveys and also signed on to a NASS letter supporting participation by their membership.[9]

DATA RELEVANCE

Contemporary Issues That Generate Farm-level Data Demands

The core requirements for ARMS data have been generated with a view of the kinds of data that decision makers and other users require in order to understand contemporary American agriculture. Understanding the U.S. agricultural sector is a daunting challenge: it has characteristics

[8]Presentation by Robert Young, American Farm Bureau Federation, September 28, 2006.
[9]Presentation by Gary Adams, National Cotton Council, September 28, 2006.

unlike other sectors of the U.S. economy, and, in part because of that uniqueness, the sector has been thought to pose special challenges in measurement and analysis.

Agriculture has long experienced a trend of increasing productivity. It has been an engine of growth and a source of net exports in international trade. Technological and structural changes have led to the largest producers becoming less distinct from other small-business sectors of the U.S. economy than in earlier years. Agriculture is becoming dramatically more concentrated. Just 15 percent of all census-defined farms produce 85 percent of agricultural product value.

Despite a shift toward greater concentration in agriculture, a large segment of the smallest producers have significant agricultural production (Economic Research Service, 2006a). Smaller family farms (smaller farms are defined as units with annual gross revenue less than $250,000) represent 91 percent of total U.S. census-defined farms, according to 2003 data, holding 71 percent of all farm assets and 70 percent of the land owned by farms. These smaller family farms make substantial contributions to the production of hay, tobacco, vegetables, cash grains and soybeans, dairy products, beef cattle, Christmas trees, and greenhouse production. Abounding in rich variety, they are classified by ERS into a typology as limited-resource, retirement, residential/lifestyle, and farming-occupation/low sales, and farming-occupation/high sales farms. Smaller farms continue to play an important role in natural resource and environmental policy, accounting, for example, for approximately 82 percent of land enrolled in the Conservation Reserve and Wetlands Reserve Programs.

Environmental and conservation issues are also high on the public agenda and generate new and complex demands for data. Large-scale producers have in many cases adopted minimum till or no-till production systems to minimize environmental impacts from runoff of agricultural fertilizers and chemicals. In general, environmental quality interacts with agricultural production, but precisely how and when are not well understood. Nonetheless, regulatory and other policy decisions have been made and are likely to increase in scope and effect. Incentives for legislatively favored management of agricultural lands have been increasingly incorporated into the annual Farm Bill. EPA regulations have impacted the livestock sector in particular in recent years. Water quality issues are a major concern throughout the United States, and agriculture plays a significant role in water pollution, whether from nutrients, pathogens, pesticides, salts, or sediment (Economic Research Service, 2006b).

Further complexity arises from the confluence of environmental and marketing issues in the production of farm products, as is evident in the increased interest on the part of producers and consumers in organic products. Related issues involve food production practices, for example, free range production of livestock, no or restricted use of genetically modified materials in some products, and stocking rates and livestock housing arrangements.

Further examples of increasing complexity and variety in the agricultural production and marketing system are such technologies as bioengineered seed, waste management practices used by livestock producers, employment of chemical and biological pest management alternatives, and information management technologies, ranging from precision farming in crop production to marketing commodities and buying inputs via the Internet. Issues involving these developments, their prevalence on smaller and larger farms, and obstacles to their productive implementation are generating demands for data, and it is natural to consider the possibilities of ARMS to supply them.

Availability of high-speed Internet access is an ongoing concern for farmers in many rural areas. Those closer to metropolitan areas or adjacent to significant population centers are much more likely to have ready access at reasonable rates. As the supply chain becomes increasingly reliant on providing information and even access to markets through Internet connections, this concern needs to be tracked with data.

Demands for information about food marketing at the consumption end are also increasing, fueled by related concerns about nutrition and its impacts on consumer health, and a recent impetus is due to evidence of a rise in obesity in the U.S. population. At the same time, hunger, food access, and poverty as it relates to food access are issues that continue to deserve attention. They are an important part of the USDA budget on an annual basis to provide food and nutrition programs through food stamps, school lunches, women and children feeding programs, and others. How household consumption behavior bears on these issues is beyond the plausible purview of ARMS, although such concerns have a bearing on what data are most worth collecting at the farm level.

Similarly, trends on the demand side influence the need for farm-level data related to exported commodities. Since the mid-1980s, suppliers of high-value products have seen export sales outpace domestic sales by a wide margin. One-third or more of fresh table grapes, dried plums, raisins, canned sweet corn, walnuts, and animal fats is exported. Nearly 60 percent of U.S. cattle hides are exported, and the export dependency of the almond industry is even higher, with 67 percent of the crop shipped overseas.

Rural community viability as it relates to production agriculture and the food processing system is an issue of continuing concern. While rural community viability relies on much more than a successful agricultural and agribusiness sector, value-added entrepreneurial activities on farms can be an important element of rural community viability.

Bioenergy is receiving heightened attention related to its potential role in rural community viability, as well as the prospects it offers for new opportunities for agricultural producers. The rapid growth in bioenergy could reshape agriculture. As energy becomes a major second market alongside food, there is interest in understanding how it may reshape the structure of agriculture. Issues that could arise include the possible impact of changes in feed markets on the structure and location of the animal population. The extent to which these new markets and new technologies will affect the structure and diversity of agriculture is also an issue that will need to be tracked with adequate data.

Role of ARMS in Meeting Data Demands

Data on farm households were published in the decennial Census of Population, as well as annually in Current Population Statistics, until 1990, but have not been since that time. Since 1990, comprehensive knowledge of the social and economic characteristics of farm households has depended primarily on information from ARMS. Similarly, ARMS provides the basic data needed for a comprehensive view of the income generated and financial health of farms of different sizes and types. There are substantial differences between the average operating profit margins and rates of returns on assets and equity for smaller and larger farms. Farm households increasingly rely on substantial nonfarm income. ARMS data provide the basic information needed to assess how farm and household economics interact for both smaller and larger farms (Economic Research Service, 2006a).

Decisions made by businesses in the agricultural and agribusiness sectors selling to and buying from farmers benefit from knowing the number and characteristics of farms in any specific target group. Data are also needed for analytical research to estimate responses by operators and farm household members to changes in markets or public policies. Public policy analysts need access to this same information to develop workable solutions that will help to further improve the technical and economic efficiency of the agricultural sector.

ARMS has at times asked respondents for information on marketing channels and contractual arrangements. The traditional markets in which prices are openly reported have small volumes, whereas contracts and vertically integrated markets are a growing trend. An issue is whether more detail should be sought on the marketing of farm products, including quality and other specifications of goods sold in addition to quantities. Information about qualitative characteristics of goods and details of production practices followed on farms is important to gain knowledge relevant to food safety, a highly visible issue in the agricultural and food sector in today's world. According to ERS, "U.S. food producers have developed an enormous capacity to track the flow of food along the supply chain, though individual systems vary" (Golan et al., 2004).

This overview of contemporary issues in the U.S. agricultural sector highlights the need for information to track trends, the types of inputs and technologies being used in the production sector, the size and relative contributions of various segments of agriculture to the overall U.S. economy, and the impacts on rural communities, keeping in mind consumer concerns and the global competitiveness of the food system today.

Private- and public-sector decision makers thus have diverse uses for ARMS data. A measure of the value of data for these uses lies in the value of the improvement in a decision as a result of having the data. In this context, it is the value of improved decisions that should drive the provision of data. The panel did not attempt to determine such values for all the uses of ARMS, but we next consider some of the most important areas of application of ARMS data.

THREE KEY AREAS

In this section, we explore in some detail the contribution of ARMS to enhancing the understanding of three specific areas of interest in agriculture policy—environmental and resource management, commodity economics, and the well-being of farming families—and we assess the adequacy of the ARMS data to illuminate these issues.

Environmental Analysis and Natural Resource Management Issues

According to USDA's estimate, there are over 2 million farms in the United States covering nearly 900 million acres of land. This land use has wide-ranging effects on the environment. ERS has produced numerous publications that draw from a variety of information sources to summarize the overall picture:

- "EPA concluded that agriculture is the leading source of pollution in 48 percent of river miles, 41 percent of lake acres (excluding the Great Lakes), and 18 percent of estuarine waters found to be water-quality impaired. . . . This makes agriculture the leading source of impairment in the Nation's rivers and lakes, and a major source of impairment in estuaries" (Ribaudo and Johansson, 2006).
- "A U.S. Geological Survey study . . . estimates that 71 percent of U.S. cropland (nearly 300 million acres) is located in watersheds where the concentration of at least one of four common surface-water contaminants (nitrate, phosphorus, fecal coliform bacteria, and suspended sediment) exceeded generally accepted instream criteria for supporting water-based recreation activities" (Ribaudo and Johansson, 2006).

- "Almost half of all wetlands in the 48 contiguous States have been drained since colonial settlement—nearly 85 percent for agricultural use" (Hansen, 2006).
- "[S]tudies suggest it may be technically possible to sequester an additional 89-318 million metric tonnes (MMT) of carbon annually on U.S. croplands and grazing lands. Based on 2001 emissions, this level of carbon sequestration would offset between 5 and 17 percent of gross U.S. Greenhouse Gas (GHG) emissions" (Lewandrowski et al., 2004).
- "Besides food and fiber, agricultural lands provide a variety of non-market outputs. These include rural amenities such as agrarian cultural heritage, open space, scenic beauty of rural landscapes, wildlife habitat, and environmental quality—all of which are unintentional byproducts of the agricultural production process" (Hellerstein et al., 2002).

Agricultural production practices result in indirect impacts, or externalities, which can degrade the environment. Transformation of undisturbed land to crop production destroys habitat for wildlife. Soil erosion, nutrient and pesticide runoff, and irrigation can pollute the air and water, degrade soil quality, and affect water supplies. The extent and degree of the environmental problems associated with agriculture vary widely across the country. Some of the key indirect impacts include:

- *Water quality impairment.* The production practices and inputs used by agriculture can result in a number of pollutants entering water resources, including sediment, nutrients, pathogens, pesticides, and salts. Assessments done for EPA by the states have found that agriculture is the leading source of impairment in the nation's rivers and lakes, as well as a major source of impairment in estuaries.
- *Water consumption.* Agriculture is the single largest consumer of water in the United States. It competes with drinking water, recreational uses, and ecosystem needs in arid regions of the country.
- *Ecosystems.* Habitat loss associated with agricultural practices on more than 400 million acres of cropland is the primary factor depressing wildlife populations in North America. Ecosystems are also adversely affected by agricultural pollutants. Nitrogen from cropland is believed to be a major contributor to the degradation of the Chesapeake Bay and to the hypoxic zone in the northern Gulf of Mexico.
- *Animal agriculture waste.* The large amounts of waste that large livestock feeding operations produce can be the source of nutrients, organic matter, and pathogens entering surface water or groundwater and gases and odors into the atmosphere.

The existence of these external effects from agricultural production and practices has generated a plethora of government programs and private initiatives to reduce the negative effects and increase the positive ones associated with agricultural production. Specifically, conservation funding by USDA has increased dramatically in the past two decades, with the Conservation Reserve Program, which pays farmers to remove land from production, representing the largest financial component. Figure 2-1 shows the total conservation expenditures by USDA over the past two decades.

In the 2002 Farm Bill, a reorientation toward increased expenditures for conservation on "working lands" was established with increased funding for the Environmental Quality Incentive Program and the introduction of the Conservation Security Program. These programs provide funds to adopt or maintain conservation practices on land that is still in crop or livestock production, thus the nomenclature of "working lands" programs.

Federal environmental laws generally adopt a voluntary approach for addressing agriculture's impacts on environmental quality. This places great importance on the design and performance of conservation programs, which rely on voluntary incentives to get farmers to adopt environmentally friendly practices. A better understanding of factors affecting program participation and performance of conservation practices would improve the overall performance of conservation programs.

These issues bear some significance for the design and implementation of ARMS. The ARMS information could play a more valuable role than it currently does through the quantification of benefits and costs of these conservation programs and the effects associated with agriculture. More specifically, two general types of analysis could be done using ARMS-type data that would meet important needs of policy makers and the public: retrospective (ex post) assessments of the costs and benefits of existing conservation programs and prospective (ex ante) studies to improve the design and cost-effectiveness of conservation programs in the future.

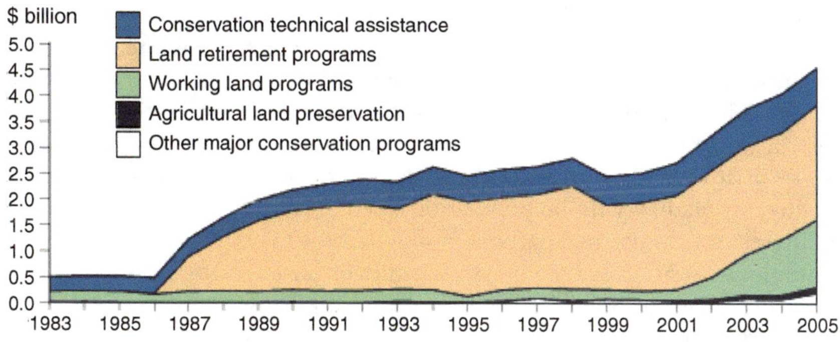

FIGURE 2-1 Trends in USDA conservation expenditures, 1983-2005
SOURCE: Analysis by USDA Economic Research Service of data from USDA Office of Budget and Program Analysis.

The substantial expenditures on conservation programs made at many levels have not escaped the notice of federal oversight agencies, such as the Office of Management and Budget and the Congressional Budget Office. In particular, USDA and other federal agencies have been under scrutiny to demonstrate the efficacy of their programs—that these large expenditures on conservation have translated into quantifiable improvements in environmental quality and lowered the costs related to these programs. Cost-benefit analyses of existing conservation programs are an example of a retrospective assessment.

A prospective study may be a cost-benefit study designed to estimate whether the benefits of a regulation or conservation program are likely to exceed the costs. Such studies may be designed to consider alternative designs of a program or regulation to improve its cost-effectiveness or assess the degree to which some other goal of the program can be met.

Both prospective and retrospective studies require estimation of the effects a conservation program has had, or is likely to have, on water quality, soil erosion, wildlife habitat, or other environmental measures of interest. Addressing this issue can be a significant challenge and requires the answer to a series of interrelated questions:

1. What changes in conservation practices and land use did these programs directly cause, in contrast to changes that might reasonably have occurred anyway? Implicit in this question is the quantification of the baseline conditions of the conservation practices and land use prior to the introduction of the program and the extent and location of the practices present after the introduction of the program—that is, an understanding of what measures were adopted in response to the specific program being evaluated. Since some conservation practices are profitable to farmers in their own right, such as conservation tillage methods with particular types of soil and climate conditions, and some farmers adopt conservation practices out of an independent desire to do so rather than as the result of a program, it is important to establish a baseline.

2. How did these changes affect the resource of interest? The second step in an evaluation requires an understanding of how these changes in land use or conservation practices affect the resource of interest: the water quality, air quality, wildlife populations, etc. Thus, to measure the endpoint of interest—a change in environmental quality—it is necessary to go beyond a simple accounting of the number of miles of buffer strips installed or conservation tillage adopted. It is necessary to link those changes to measures of in-stream water quality, songbird populations, or carbon sequestered. Most efforts along these lines are done via biophysical model predictions rather than direct monitoring.

3. What did these changes cost? This should include both the direct

costs of installing practices, such as buffer strips or terraces, and indirect costs, including lost revenues from higher input costs or lower yields, the opportunity cost of time associated with managing conservation activities (such as nutrient management, for example), and the higher economic risk that is attributed to some conservation activities.

The data needed to undertake small-scale or national assessments of the effectiveness of a program include the following:

1. Detailed information on the physical resources of a farm: its soils, weather characteristics, etc.;
2. Information on preprogram land use: cropping patterns, rotations, conservation practices, fertilizer and pesticide usage, etc.;
3. Information on changes in behavior induced by the program and the full costs associated with those changes;
4. Data on payments or compensation received by farmers as a result of program participation; and
5. Data on effects on the resource of interest (soil erosion, off-site water quality changes, etc.).

The ARMS data currently collected contain some of the information useful for national or regional assessments of the sort just described, but they fall short at several levels. To do a complete retrospective analysis, it is necessary to also link the land characteristics (soils, proximity to water bodies, slope, etc.) with land use decisions (cropping systems, conservation practices, etc.), farm costs and returns, and participation in conservation programs. Each of the first three linkages is needed for prospective studies.

NRCS houses the National Resource Inventory (NRI), which is conducted in cooperation with the Center for Survey Statistics at Iowa State University and provides information that assists development of agricultural-environmental policies and programs. The NRI was designed and implemented to assess conditions and trends in soil, water, and related resources on nonfederal rural lands. Accordingly, it captures data on land cover and use, soils, soil erosion, wetlands, habitat diversity, selected conservation practices, and related resource attributes. The NRI sampling procedure employs aerial photos of land use and changes in use. The design includes 300,000 area segments (about a half-mile square area) and 800,000 points within the segments (about 3 points per segment). These NRI points are located across the nation and are monitored at periodic intervals by aerial photo interpretation and on-site visits. This longitudinal survey yields time series data for the 800,000 sample sites and has produced consistent historical data back to 1982, converting from a 5-year cycle to an annual inventory in 1997.

While the NRI provides information about resource conditions and environmental outcomes, it lacks equally important information about production and economic effects. Linking information on farm costs and practices directly with the resource characteristics of the farms is key to assessing retrospective benefits and costs of existing conservation programs.

ERS uses ARMS data for conservation policy analysis, including questions on what types of farms participate in USDA conservation programs, how much conservation and commodity program participation overlap, whether conservation compliance reduce soil erosion on U.S. cropland, and if conservation tillage is good business as well as good stewardship (Hopkins and Johansen, 2004).

One limitation of ARMS in conservation policy analysis is that the information available in the survey on conservation program participation is very limited. ARMS has little to offer in terms of resource conditions or potential environmental outcomes, both of which are critical to conservation policy analysis. A partial remedy would be to revise ARMS questions to ask about conservation program contracts, rather than payments, so that the answers would fill gaps in administrative data. Program management could be improved by determining who applied for conservation programs but was rejected and who has adopted conservation practices without the benefit of conservation program payments. Similarly, the use of county- or watershed-level proxies in ARMS misses a lot of variability across finer classifications of land areas, compared with proxies that could be measured on-site.

Resource conditions—such as the vulnerability of soils to erosion or nutrient runoff and leaching and location of the land relative to streams or rivers—are essential to conservation policy analysis. ARMS provides very limited information about resource conditions. The Phase II survey has information about highly erodible land and wetlands status, but it is limited to specific crops in specific years.

Another major shortcoming of ARMS data is the lack of precise geospatial information. Accurate geocodes are key to linking resource data with ARMS Phase II, but current geocodes give locations that are, on average, two miles from the survey field.[10] Accurate geocodes derived from a geographic positioning system would facilitate links to extensive geographic information system (GIS) data on resource quality, climate, urban influence, and the like. The collection of accurate geocodes in the ARMS program has been tested using a global positioning system (GPS) with good results (Claassen, 2006; see Chapter 5).

USDA administrative data could be linked to ARMS to provide useful

[10]Phase III provides only general location in the form of county and zip code.

information for conservation policy analysis. Although ARMS data are useful in conservation policy analysis, gaps remain on resource conditions and conservation program participation. Linking administrative data, using accurate geocodes on Phase II for linking GIS-based resource data, and changes to ARMS questionnaires could partially fill these gaps.[11]

Another attribute of survey data that needs to be considered when assessing its suitability for either prospective or retrospective studies is whether adequate sample size is available to characterize the relevant population. For example, if one is interested in assessing the costs and soil erosion consequences of a particular set of payment options in the Conservation Security Program, it would be natural to identify a watershed as the appropriate population, since the program is implemented watershed by watershed.

Commodity-Specific Practices, Costs, and Returns

Commodity-specific information is collected on a rotating basis in both the field-level (Phase II) and whole-farm (Phase III) portions of ARMS. Production practice data for major crop and livestock activities (corn, soybeans, wheat, cotton, dairy, and hogs) are gathered more often than those for other commodities (other feed grains, other small grains, sugar beets, rice, peanuts, tobacco, and poultry). Livestock data (cow-calf, hogs, and dairy) have been collected approximately every five years on a staggered rotation (Table 2-1). The selection of crops may appear to be somewhat haphazard, but there are forces at work that define coverage. Crops are selected based on multiple factors, including emerging issues and changes in the farm program impacting various commodities. Also, there is an attempt to coordinate collections to NASS environmental and other program rotations to reduce respondent burden. Budget and the overall sample size limits for the survey serve to constrain the number and type of crops that can be surveyed in a given year.

The Commodity Cost and Return (CAR) estimation project at ERS is based on commodity surveys (Phases II and III for crops; Phase III for livestock) from ARMS. These estimates are available by commodity on the ARMS website, and the CAR page is one of the most accessed pages on the ERS site (2,000-2,500 unique hits per month). The CAR data support research projects both at ERS and at universities.[12]

The collection of data on production practices and costs is much more comprehensive and leaves fewer unexamined topics than is the case for the resource and environmental issues discussed in the preceding section. This is a natural outcome of the long prehistory of ARMS in USDA's cost-of-production

---

[11]Presentation by Roger Claasen, Resource and Rural Economic Division, Economic Research Service, September 28, 2006.

[12]Presentation by Bill McBride, Economic Research Service, September 28, 2006.

TABLE 2-1 ARMS Coverage by Commodity and Year

Commodity	1996	1997	1998	1999	2000	2001	2002	2003	2004	2005	2006
Corn	B	A	A	A	A	B				B	
Soybeans	A	B	A	A	A		B				B
Cotton	A	B	A	A	A			B			
Winter wheat	A	A	B		A				B		
Spring wheat	A	A	B		A				B		
Durum wheat	A	A	B		A				B		
Fall potatoes	A	A		A							
Rice					B						B
Sorghum (milo)								B			
Flue-cured tobacco	B										
Sugar beets					B						
Peanuts				A					B		
Sunflowers				A							
Oats										B	
Barley								B			
Cow-calf	C										
Hogs			C							C	
Dairy					C				C		

A = Phase II field-level Production Practices Report only.
B = Both Phase II field-level Production Practices Report and Phase III whole-farm Cost-of-Production Survey.
C = Phase III whole-farm Cost-of-Production Survey only.

estimation program. Expansion of questions from production costs to areas of farm product marketing have added much to the knowledge of these farm activities, but the survey approaches have not yet become standardized as they have for questions covering production costs. Even for production costs, ARMS still does not provide a complete picture of farm economics. The fact that each year's outcomes with respect to both outputs and prices are subject to substantial randomness because of weather and market conditions, among other factors, means one can never be confident that a single year's survey results are representative of that farm, or that the distribution of results across all farms is representative of the structure of farm output or costs across types, locations, or sizes of farms. A panel approach to ARMS would be one way to address this limitation (see discussion below).

The vagaries of weather and market conditions have a significant effect on the cost and returns structure of individual farms, and these must be considered in the long term, not just at a point in time. A Canadian study based on longitudinal data (Culver et al., 1991) showed that the average farm (regardless of size) in a four-year panel of grain farms returned about 26 cents of cash revenue per dollar of gross revenue. The one-quarter least efficient farms broke even ($0.00 cash revenue per dollar of gross revenue) over the four-year period (regardless of farm size), and the one-quarter most efficient farms generated over 50 cents of cash revenue per dollar of gross revenue (regardless of farm size). The systematic difference in the level of efficiency was dramatic. Using longitudinal data, it was possible to identify the farms that were systematically efficient and those that were systematically inefficient. The observation of these characteristics at one point in time cannot indicate whether the results are transitory or structural. The promise of longitudinal data for this type of analysis leads to the discussion of the utility of longitudinal data at the end of this chapter.

Farm Household Economics

Surveys collecting data on household income and well-being have progressed over the years to more completely account for the dynamic economy, and agricultural surveys are no exception. The structure of the American farm continues to evolve, with the delineation between family and nonfamily farms becoming more important, as is the increasing value of off-farm income in measuring farm household well-being. The comparison between the farm and nonfarm populations is complicated by the complexity of determining farm income and wealth and the different sampling frames and geographic coverage of such agricultural surveys as ARMS and such household surveys as the Current Population Survey.

In the Soil Conservation and Domestic Allotment Act of 1936, Congress declared its purpose as the reestablishment of "the ratio between the purchasing power of the net income per person on farms and the income per person not on farms that prevailed during the five-year period August 1909-July 1914" (U.S. Department of Agriculture, 1944). The lack of income parity between the farm and nonfarm population was a central component of the "farm problem" as defined at that time.

There were several early sources of statistics on the income of farm households. The first statistical series compared the disposable personal income per capita of farm residents with that of nonfarm residents for the years 1910-1943. The goal of this series was to build on USDA's aggregated sector-level estimate of net farm income to get farm income per farm, by dividing by the number of farms in the Census of Agriculture and supplementing the resulting estimate of farm income per farm with information on off-farm income available occasionally from the Census of Agriculture and other sources. USDA published this derived series until 1983.

A later historical series, on the incomes of principal farm operator households, began in 1960. This series was the sum of the annual estimate of the net farm income of the sector and off-farm income of the farm operator household based on Census of Agriculture data. A major deficiency of this series was inconsistency in the treatment of cash and noncash income items, and it was replaced by one that treated income items in a way consistent with the treatment of the income statistics for U.S. households based on the Current Population Survey. For 1984 and later years, estimates were based on the Farm Costs and Returns Survey (FCRS) data. The series was later refined with the 1988 FCRS data in a variety of ways, including recognizing that not all farms are family farms and that not all farm business income went to the farm operator household. This series is the basis of the current ERS statistical series on farm operator household income and is compared with the incomes of the average U.S. household and published annually.

This is not a straightforward matter, since comparing farm household income from ARMS with the income of the average U.S. household from the Current Population Survey (CPS) requires calculating an estimate of farm household income from ARMS that is consistent with CPS methodology. The CPS defines farm self-employment income as net money income from the operation of a farm by a person on his or her own account, as an owner or renter, and includes income received as cash but excludes in-kind or nonmoney receipts. Farm self-employment income from ARMS is the sum of the operator household's share of farm business income (net cash farm income less depreciation), wages paid to the operator, net rental income from renting farmland, and other farm-related income (net income from a farm business other than the one being surveyed, wages paid by the farm business to household members other than the operator, and commodities paid to household members for farm work) (Economic Research Service 2005).

Although USDA's estimates of income from farming are based on survey data, they have an important feature that is not followed in estimating nonfarm self-employment income of households: farm operators are not asked directly about their net income from farming. Instead, ARMS collects data about the various components of farm income—revenues from selling products and services and costs of production. ERS then calculates each farm's net income in several ways, using cost concepts that fit with different income concepts. For example, cash income does not account for depreciation of capital, while net farm income does. It is important to provide different measures of farm income because the use of data derived directly from survey responses varies among them.

To obtain an estimate of a farm household's off-farm income, ARMS does ask directly about net income received, even for off-farm business income that may have returns and costs complications similar to those of farming.[13] The data show a striking trend of farm operator household income overtaking the incomes of nonfarm households over the past two decades.

The collection of farm income data on ARMS is an important input to the USDA sector-wide set of income accounts. The cash revenue received in the sector is estimated, generally, by multiplying an average price times an estimate of the quantity marketed. Data on the direct government payments to the sector come from the government agency making the payments. In general, the data on cash expenses come from the ARMS program. There are exceptions—for example, the estimate of interest expenses comes from agencies that make loans to the agricultural sector. The depreciation is calculated based on the value of buildings and the value of machinery and equipment reported by the ARMS program.

One contribution of the ARMS microdatabase is the ability to disaggregate the sector-wide estimates of farm income in order to show the flows to and from each of the major stakeholders in the agricultural sector. To do this requires detailed enumeration by the ARMS program. Specifically, the ARMS questionnaires determines the revenue flows and expenditure flows for the following groups:

- Family farms—generally, these are proprietorship farms, partnership farms, and incorporated farms operated by family members.
- Nonfamily farms—generally, these are farms operated by widely held corporations.
- Landlords—generally, these are individuals or companies that rent land to farmers—but sometimes they lease cows or machinery to farmers—and often they contribute part of the chemical and fertilizer expenditure for the operation of the land that they rent to their tenant.

[13]The 2005 Core Phase III survey asks for "the principal operator's pay from operating any other business" and for "the cash income the principal operator, spouse, and all other household members received in 2005 from" a list of categories including "other off-farm sources of income," which is where business income other than the principal operator's pay from such business would be reported.

- Contractors—generally, these are businesses that establish contracts with farming operations in order to participate (or "control") some or all of the decision making in crop or livestock production. For example, in egg production, the contractor may own the birds and pay the farm operator a management fee to compensate for the feed, the barn, and the labor. Alternatively, the operator may own the birds but the contract specifies the number, quality, and quantity of eggs to be delivered. There is a wide range in the types of contracts and the types of incentive schemes in these contractual relationships. Importantly for the purpose of analysis, these businesses ("contractors") receive part of the return to labor and capital (i.e., the net farm income) in the sector.

The task of determining the size of the various flows associated with each enumerated farm adds "enumeration complexity" to an already complex survey environment. Nonetheless, as a result of collecting these flows, USDA can show the role of each stakeholder group. Specifically, for 2005, estimates provided by ERS indicated:

- $74 billion aggregate sector-wide net farm income; of which
    - $49 billion was received by family farms
    - $9 billion was received by nonfamily farms
    - $2 billion was received by landlords
    - $16 billion was received by contractors
    - $2 billion was discrepancy

The panel requested documentation of the discrepancy, which requires detailed analysis to reconcile (see Box 2-1). ARMS collects cash revenue for each observation, and the sum across all observations does not replicate, exactly, the results of the methodology for estimates of cash revenue for the sector-wide accounts (described above). In addition, there were numerous other sources of discrepancies—this is not surprising if one is generating statistics from different sources.

The panel found the detailed reconciliation to be instructive as an indicator of the size of the differences in the estimates from the ARMS program and the published sector-wide estimates (Box 2-1). Making this reconciliation available to users on an annual basis would enable them to understand the different sources of data in the different accounts and to appreciate the magnitude (or lack thereof) of the discrepancies. In the final analysis, significant discrepancies exist between the expenditure estimates and the income estimates in the national accounting framework for estimates of gross domestic product.

In addition to providing a disaggregation by type of stakeholder in U.S. agriculture, ARMS calculates a farm-level estimate of net value-added,

BOX 2-1
Reconciling the Apparent Gap in Net Farm Income and Farm Household Earnings

Aggregate net farm income for the agricultural sector was $82.5 billion in 2004.

"Earnings of the operator household from farming activities" was $14,201 per household in 2004. Assuming 2,060,822 "family farms," the aggregate calculates to be about $29.3 billion.

ISSUE: How to get from the $82.5 billion for the agricultural sector to the $29.3 billion for "family farms."

POSSIBLE EXPLANATIONS: The net farm income line of Table 2-2 for 2004 provides the main elements of the gap as estimated by ERS. They are net farm income received by landlords who are not farmers, nonfamily farms (notably corporations in farming), and contractors who pay farmers to produce commodities but are not themselves farmers.

RECONCILIATION: In order to compare aggregate sector-wide estimates of net farm income with an estimate of farm operator household income from farm and nonfarm sources, several adjustments must be made to the estimate of aggregate sector net income. These include (1) distributing aggregate sector-wide income to stakeholders, as depicted in Table 2-2; (2) the remaining net farm income of operators, which is a measure of profit earned, must then be converted to an estimate of net cash income; (3) the estimate of net cash income of operators must be distributed to households that participate in the business operation (many farms, particularly, but not exclusively, larger businesses, have multiple operators); and (4) the business measure of net cash income that accrues to the primary operator must be converted to a census-based measure of money income. All these steps result in an estimate of "income from farming activities that accrues to the principle operator households" not aligning with aggregate sector-wide estimates of net income, even when the per farm estimate is expanded to account for one principal operator household for each farm.

SOURCE: T. Covey et al., 2005, p. 1.

which allows the tabulation of value-added by such variables as type of farm, size of farm, and geographic area (Table 2-2).

The approach to farm income measurement is predicated on the selection of an appropriate statistical indicator for several possible measurements of interest. In a presentation to the panel, ERS provided the information shown in Box 2-2, which ties the appropriate statistical indicator for farm income to the concept to be measured.[14]

[14]Presentation by James Johnson, Economic Research Service, September 28, 2006.

TABLE 2-2 Net Income Derived for ARMS Using Sector-wide Account Procedures, 2004 and 2005 (dollars in billions)

	2004					
	Operators					
	Family Farms	Nonfamily Farms	Landlords	Contractors	ARMS Total	Estimates Sector-wide Income Accounts
Value of crop production	91.5	14.6	5.9	1.3	113.0	125
Value of livestock production	69.6	13.7	0.0	39.7	123..0	124
Revenues from services and forestry	21.1	1.1	11.4	0.0	33.7	34
Value of agricultural sector production	182.2	29.5	17.4	41.0	270.0	283
Purchased inputs	96.6	14.8	1.2	26.2	138.8	137
Farm origin	34.6	8.2	0.3	19.6	62.7	58
Manufactured inputs	28.6	2.5	0.7		31.9	32
Other intermediate expenses	33.5	4.0	0.3	6.6	44.3	48
Net government transaction	4.6	-0.1	-0.7		3.9	5
Farm gross value-added	90.2	14.6	15.5	14.8	135.1	152
Capital consumption	16.3	1.3	2.8		20.4	23
Farm net value-added	73.9	13.3	12.7	14.8	114.7	129
Payments to stakeholders	32.7	5.7			38.5	43
Net farm income	51.5	8.2	1.8	14.8	76.2	85

TABLE 2-2 Continued

	2005					
	Operators					
	Family Farms	Nonfamily Farms	Landlords	Contractors	ARMS Total	Estimates Sector-wide Income Accounts
Value of crop production	83.8	17.2	5.0	1.8	108.0	113
Value of livestock production	71.5	11.1	0.3	38.4	121.2	127
Revenues from services and forestry	22.0	1.8	12.3	0.0	36.1	36
Value of agricultural sector production	177.2	30.1	17.7	40.2	265.2	275
Purchased inputs	102.0	14.5	1.4	24.6	142.6	147
Farm origin	32.4	5.6	0.4	15.7	54.1	58
Manufactured inputs	31.7	3.2	0.7	0.2	35.8	36
Other intermediate expenses	37.9	5.7	0.3	8.7	52.7	53
Net government transaction	10.0	0.4	-1.8		8.6	16
Farm gross value-added	85.2	16.0	14.4	15.6	131.2	145
Capital consumption	15.6	1.3	2.7		19.5	24
Farm net value-added	69.7	14.7	11.7	15.6	111.7	120
Payments to stakeholders	32.3	7.3			39.6	46
Net farm income	46.7	8.4	1.8	15.6	72.0	74

SOURCE: James Johnson, *Reconciling Sector-wide Net Farm Income and Farm Household Income Data Series*, ERS, September 28, 2006.

BOX 2-2
Income Measures and Statistical Indicators

Measure	Statistical Indicator
Farming's contribution to state and national economies	Farm sector's net value-added
Earnings of farming's risk takers	Sector-wide net income
Net income of farm business establishments	Farm business income
Income earned by farm households	Money income of farm households from nonfarm and farm sources
Income available for household use	Farm household's disposable (after tax) income

ARMS data on farm income are an important contributor to the national income and product accounts. In fact, the need to populate the national accounts has driven decisions on the content of the ARMS program over the years. The content of the questionnaires has been written to generate data consistent with the accounting requirements of the national accounts.[15] The data collected for this purpose include production expenses, imputed rents, sources of farm-related earnings other than the value of production and receipts, capital items such as farm office equipment, and data by the organizational structure of farms that can be used to distribute wages and other items by legal form of business organization.

Beyond the content of questionnaires asking explicit questions for use in the farm income accounts, ARMS produces a farm-level estimate of value-added that is as consistent as possible with sector-wide measures of net value-added and farm income. Weighted estimates of firm-level value-added can be compared with sector-wide estimates produced from multiple sources of data as a consistency check. The firm-level data can be disaggregated to show where value-added and net income were generated by type and size of operation and geographic area.

Other means of comparison are available. Data edit, analysis, and calibration programs have been written that enhance the ability to identify and

[15]The questions that are used for the national accounts include Sections A, B, and C of Phase III (data on the whole picture of the farm), Sections D and E (income flows), Section F (expenses), and Section G (management). Sections I and J are used to fill in areas of farm and operator characteristics.

check potentially problematic responses while the survey is ongoing. The calibration routines result in ARMS data matching official USDA estimates for farm numbers by size of operation and for crop acreages and poultry and livestock on farms. ARMS has been expanded to generate directly observed estimates of farm expenditures for the 15 largest states as measured by volume of farm sales—called the 15-state oversample. For those states (Arkansas, California, Florida, Georgia, Illinois, Indiana, Iowa, Kansas, Minnesota, Missouri, Nebraska, North Carolina, Texas, Washington, and Wisconsin), directly observed annual data can be used in establishing estimates that feed into the national accounts.

Closely associated with issues of farm income are issues of the well-being of farm families. Farm household well-being, however, is a more comprehensive measure than income, involving off-farm income, which depends on family composition, wealth, labor force participation, and other factors. The concepts and measures used to determine well-being were designed for comparability with Census of Population and other major survey definitions. The primary measure of household income is money after-tax income.

In comparing the farm household measures from ARMS with those calculated from Current Population Survey, the two areas of comparison are population and the actual income measures. Since 50 percent of areas with farms are not in the CPS (primary sampling unit) sample, there are issues of sampling in these comparisons. In addition to comparisons with CPS data, ARMS measures have been compared with the Survey of Consumer Finances and Internal Revenue Service (IRS) data. The comparison with IRS is complicated by the difference in populations covered, the conceptually different income measures, and the fact that IRS has documented substantial underreporting of farm income on tax records.

There are several peculiarities in the ARMS data related to measuring facets of well-being. Some of the survey questions require considerable recall; for example, the questions on household living expenses ask respondents to recall, in the winter of the reporting year, expenses incurred over the prior year. The point-in-time reporting of expenses may result in misleading measures of facets of well-being. Expenses are episodic, subject to seasonal fluctuations, so annual estimates cannot be calculated by multiplying the reference week by 52. These issues are grist for consideration in a formal methodological research and development program, which we recommend in Chapter 3.

MAINTAINING RELEVANCE BY DEVELOPING NEW DATA SOURCES

The above discussion indicates the value of collecting conceptually correct data for individual farms and households in order to understand the structure of returns in U.S. agriculture. The issue of returns is illuminated by the currently integrated ARMS program. The success of its pioneering integration of data collection about farms and farmers prompts exploration of further opportunities for integrated data collection.

Conservation Effects and Assessment Project

A recent attempt to produce an integrated database in order to quantify the environmental effects and benefits of conservation programs is the Conservation Effects Assessment Project (CEAP), sponsored by the National Resources Conservation Service. CEAP has two main components: (1) watershed case studies; and (2) national assessments of cropland, grazing lands, wetlands, and wildlife. The goal of the program is to provide farming communities, the public, legislators, and other stakeholders with an interest in environmental policy issues with an accounting of the environmental benefits obtained from these conservation programs. By design, the CEAP cropland survey borrows heavily from the ARMS program.

To conduct the survey, a subsample of NRI points, as described earlier in this chapter, is selected from points that are classified as cultivated cropland in the most recent annual NRI survey. These points numbered 10,000 in 2003, 2004, and 2005 and were cut to 6,000 in 2006. The NASS interviewers fan out to interview operators of the farm or field associated with NRI sample point. The CEAP database is enriched with the addition of USDA Field Office records, which are used to identify program participation and practice application.[16] This is a unique aspect of CEAP, in that these records are not currently collected in the ARMS program.

The CEAP farmer survey overlaps considerably with Phase II of ARMS and could be considered in competition with that survey operation. Like the ARMS Phase II survey, it is administered late in the year, after harvest, by NASS. It asks questions going back three years on cropping patterns/rotations, field characteristics, double cropping, cover crops, and soil test results; applications of fertilizer, manure, pesticides; timing and equipment used for all field operations; tillage, cultivation, chemical application, and harvest; conservation practices associated with the field; from the respondents and the Field Offices, conservation plan and program participation; general characteristics of farming operation and the operator; irrigation practices; and some subjective information on wildlife.

While NRCS plans to continue the analytical aspect of the CEAP with the data generated through the program, no further CEAP survey has been planned at the time of this writing. A decision has been made to forgo the farmer surveys in 2007 and to look at the potential of using NRI, CEAP, weather, and other data in an integrated fashion to paint a picture of the

[16]Presentation by Jeff Goebel, September 28, 2006.

environment and its impact on soil quality and water quality and quantity, as well as on-site and off-site environmental impacts; the idea is that this approach would become a standard part of the NRI survey process after 2007 for grazing lands, wetlands, and cropland. The "time-out" affords an opportunity for considering whether it makes sense for NRCS, NASS, and ERS to work on an integrated approach for surveying agricultural lands. In the panel's view, there are serious unanswered questions about survey methodology and cost-effectiveness that need to be answered before a decision can be made that some integrated effort should be launched.

Recommendation 2.1: NRCS, NASS, and ERS should engage in a focused research and testing program and use experience with integrating the Conservation Effects Assessment Project and ARMS to assess the feasibility of integrating ARMS with other surveys and data sources.

Longitudinal Data from ARMS

In 2006, ERS and NASS submitted a proposal for funding to begin to develop the Agricultural and Rural Development Information System (ARDIS). ARDIS has several objectives, one of which is to establish and maintain data collection on the demographic characteristics, employment, and income sources of rural households over time. ARDIS would collect information on participation in farm, rural development, and other USDA and federal programs, enriching it with household well-being information, in a new longitudinal survey of nonfarm rural households and rural-based farm households.

The agencies envisioned that ARDIS would allow researchers to isolate the effects of rural development, farm conservation, and marketing programs from one another and from the myriad other forces affecting the economic well-being of farm and rural households. The close interaction among farm and nonfarm rural households is critical to understanding how rural America adjusts to changing economic circumstances or policy over time. For example, farm operators who report receipt of government payments in one year could be followed to determine the effect on management practices and income in subsequent time periods, as a means of evaluating the effectiveness of these programs at the individual farm level. The unique, previously nonexistent, longitudinal ARDIS data would help discern these patterns of adjustment.

As proposed, ARDIS would be a simultaneous collection of ARMS data and information from separate panels of farm and nonfarm households over time in the same rural area. The goal is to gauge how farm and nonfarm rural households respond differently to economic change and to enhance the ability to assess the impact of farm policy on the rural economy. The new program would track critical indicators over time with special-focus modules to address specific, emerging policy issues; monitor farm and rural household adjustments to changes in policy, prices, and technologies by building on ARMS; and enhance research capacity to analyze, interpret, and apply new agricultural and rural development information.

At the time this report was prepared, the funding for development and implementation for ARDIS was not available to the agencies. This initiative bears watching as it would provide a unique data source for enriching analysis of the interplay between programs and performance results.

Recommendation 2.2: In preparation for funds becoming available for a longitudinal design of ARMS, ERS and NASS should systematically conduct research and explore the need for and feasibility of obtaining panel data from ARMS. Furthermore, as a test of the power of such information, more use should be made of the existing longitudinal microdata from the Census of Agriculture. One possible approach would be to create a pseudopanel of such data. Another would be to make a retrospective link between the Census of Agriculture and a year of ARMS.

3

Survey Management

The complexity of the Agricultural Resources Management Survey (ARMS) extends to matters of direction, administration, and financing. Although this situation is by no means unique in the federal statistical system, understanding it does require a thorough understanding of the substantive, statistical, and institutional dynamics underlying the survey. A pattern of governance has evolved that includes a steering committee and numerous supporting committees and boards. This shared web of governance has enabled the survey to move forward gradually, but it may require reorganization in the future, depending on the directions for future growth chosen by the U.S. Department of Agriculture (USDA) or Congress.

COLLABORATIVE MANAGEMENT

Representing USDA in managing ARMS, the National Agricultural Statistics Service (NASS) and the Economic Research Service (ERS) share responsibility for the subject matter in a complementary manner. The approaches the agencies take to the various management tasks are quite different, because the roles and missions of the agencies are different.

NASS has a dual role: it is both the data collector and a major user of data derived from ARMS. The statistics it produces are usually seen as descriptive. They are for the most part, univariate counts, means, and totals for specific classifications of farms, farm production, or crops. Data on agricultural commodities, production costs and expenses, chemical and pesticide use, and farm income and assets find their way in tabular form into NASS publications and other agency products.

The role of ERS is less sharply drawn. At one level, ERS can be characterized as a consumer. Indeed, if a distinction were drawn between producer and consumer of the ARMS data within USDA, it would be appropriate, if not entirely accurate, to identify NASS as the producer and ERS as the consumer. But ERS is a very active and involved consumer. Because ERS considers ARMS to be vital to accomplishing its mission of research and analysis, the agency plays a significant role in driving ARMS content, particularly in areas that call for knowledge of the economics of agricultural production and the farm household.

Like NASS, ERS produces descriptive statistics, although its primary focus is analytical. To support its policy work and longer term research, ERS makes extensive use of multivariate models. A fundamental difference between descriptive and analytical work is that the former can often be computed using independent sources of information, whereas the latter typically requires the full set of variables collected from each unit observed. Thus, all other things being equal, analytical users tend to press for increasing the scope of a survey, while descriptive users may be more sensitive to respondent burden issues that may lead to nonresponse or other aspects of survey operations that would contribute to error and variability.

The influence of ERS is most strongly present in the development and analysis of the data from the Phase II and Phase III surveys. ERS has expertise in the development of information on environmental resource management and has worked collaboratively with NASS to frame the Phase II collection of data on chemical and pesticide use on cropland. The agency's primary interest and ownership is over the Phase III survey operations, in which ARMS collects basic economic data on income, expenses, and debt annually.

The ARMS economic questionnaire supports an ERS program of data analysis on farmers' use of particular marketing channels and on management decisions and farm household well-being, including operator demographics. By combining data from the Phase II and Phase III questionnaires for the overlapping portion of the sample, ERS is able to add value by making ARMS a very powerful survey for analyzing the relationship of the environmental and economic components of agricultural production.

Program Funding

Most of the funding for ARMS has traditionally come from NASS, although ERS contributes substantially and increasingly to the financing of the survey. The reimbursement from ERS totaled $6.75 million for the base survey in fiscal year (FY) 2005 and FY 2006. In FY 2005 and 2006, NASS funding continued at the FY 2003 level of $9.9 million. The cost of ARMS

TABLE 3-1 ARMS Program Funding by Agency, Fiscal Years 1996-2006 (in millions of dollars)

Fiscal Year	National Agricultural Statistics Service	Economic Research Service	Total
1996	N/A	$2.3	
1997	N/A	4.0	
1998	N/A	4.0	
1999	N/A	3.9	
2000	N/A	3.7	
2001	$5.1	3.6	$ 8.7
2002	5.3	4.0	9.3
2003	9.9	6.4	16.3
2004	9.9	6.7	16.6
2005	9.9	6.7	16.6
2006	9.9	6.7	18.7

N/A = Not Available.

was approximately $18.7 million in FY 2006, not including the time of about 36 ERS staff who are actively involved on at least a part-time basis with ARMS data development or who use ARMS data in their research.

The funding for the survey jumped significantly in FY 2004, when an additional sample was added to provide data for a 15-state oversample, going from less than $10 million per year to over $16 million. Since that quantum jump in resources, the ERS contribution has been accelerating and the NASS contribution has remained fairly constant, although NASS continues to provide the bulk of the funding (Table 3-1).[1]

The FY 2005 NASS survey costs were distributed between data collection (44.7 percent), staff (48.2 percent), and direct and indirect (7 percent) costs.

The ERS resources committed from its general appropriation to ARMS data development and research involve several units in the following program areas: agricultural structure and productivity, farm and rural household well-being, farm and rural business, and production economics and technology. Depending on the subject matter of interest, four other program areas have a role on the NASS/ERS steering committee: diet, safety, and health; animal products; grains and oilseeds, specialty and fiber crops; and resources, environmental, and science policy. About 7.5 full-time-equivalent employees are dedicated to ARMS, and 20 additional researchers in ERS use ARMS data. The full-time staff for ARMS is concentrated in the Resource and Rural Economics Division.

[1]ERS funding for FY 2004 and FY 2005, includes $200,000 and $250,000, respectively, to fund collection of data on organic commodities—in 2005 for Phase III organic dairy and, in 2006, for Phases II and III organic soybeans.

Division of Responsibility for ARMS Functions

As it has evolved over time, the division of responsibility for ARMS follows the lines described above for each of the key functions of survey management: design, development, operations, and analysis/dissemination. Table 3-2 summarizes these functions.

TABLE 3-2 Division of Responsibility for ARMS by Function

Key ARMS Functions	National Agricultural Statistics Service	Economic Research Service
Resources	Obtains base resources for program of statistical methodology, survey operations, and preparation of NASS publications.	Obtains supplemental resources to provide to NASS for conducting survey and funding for organics research and internal ERS funding for analysis and dissemination program.
Survey approval process	Prepares and defends OMB information collection request for survey.	Assists in preparation of OMB information collection request.
Identification of concepts and desired outputs	Develops and implements statistical concepts.	Aligns survey topics and questions with measurements and analytical goals.
Survey specifications	Obtains feedback from state offices, data collectors, and respondents regarding past surveys and develops questions based on needs of internal and external users.	Identifies new or revised questions to be included in the survey based on interactions with users.
Cognitive testing of questionnaires and modes of collection research	Conducts program of cognitive development and testing of questionnaires and research into modes of collection.	
Presurvey activities	Identifies resources and develops operating plan for upcoming survey.	
Sample design	Primary responsibility for designing the surveys and integrating the phases.	Reviews the sample allocation to offer analytical insight into state and commodity coverage.
Edit design process	Establishes criteria and designs the edit.	Offers input into edit parameters.
Preparation of manuals and training materials	Prepares the survey operations manuals and training materials.	Participates in writing the sections of enumerator manuals that pertain to analytical issues.
Conducting supervisor and enumerator training	Conducts the field supervisor (survey statistician) and the state-level enumerator training in coordination with NASDA.	Participates in survey statistician training; assists in conducting state-level training.

continued

TABLE 3-2 Continued

Key ARMS Functions	National Agricultural Statistics Service	Economic Research Service
Predata analysis	Develops data analysis tools that will be used by state and national state during the survey process.	Contributes expertise in development of data analysis tools.
Data analysis	Produces a clean data set with as few data "errors" as possible. Uses computer interactive data analysis capabilities.	Assists in use of computer-interactive data analysis; assists as needed and depending on the availability of staff.
Imputation and final edit	Conducts imputation and edit functions.	Assists on a periodic basis.
Summary	Prepares summary data files.	
Outlier identification and board process	Takes lead in interacting with states to draw input on possible outlier records; convenes and manages outlier board; develops rules for outlier identification.	Reviews information and participates as members of outlier board; assists in developing rules for outlier identification.
Final summary	Prepares final summary data files.	
Transmittal to ERS	Transmits data files to ERS.	
Farm production expenditures board	Conducts production expenditures board, regional review board, and state review board.	Participates in the three board processes.
Releases of farm production expenditures	Publishes farm production expenditures release in paper format and electronic format via the Internet.	
Releases of other Phase II and Phase III information		Publishes several products to release Phase II and Phase III data to public.
Preparation of research databases		Uses ARMS file to produce estimates of farm business and household income and balance sheets for release through briefing rooms and ARMS data tool; prepares ARMS research databases.
ARMS data dissemination tools	Participates in developing ARMS data dissemination tools to ensure that appropriate rules are followed for maintaining data confidentiality.	Develops and maintains ARMS data dissemination tools.

OMB = U.S. Office of Management and Budget; NASDA = National Association of State Departments of Agriculture.

Coordinating Mechanisms

The NASS/ERS Steering Committee is the principal coordinating mechanism for ARMS. This committee is co-lead by the environmental and economic surveys section head from NASS and the ERS Deputy Director for Resource and Rural Economics Data, and is composed of senior management staff in each division of both organizations—six from ERS and four or five from NASS. There is a lot of interaction between the agencies, their various branches, and the steering committee.

NASS and its branches focus on some of the following topics in the steering committee:

- The environmental, economics, and demographics branch in the statistics division serves as coordinator between NASS and ERS in matters of analysis design, and implementation, data analysis strategies, coordinates outlier treatment, and is responsible for the estimation process that culminates in NASS publications and delivery of the final data set to ERS. The economics section has responsibility to develop and implement specifications for all edit, summary, analysis, and estimation programs.
- The sampling branch designs the list and area frame samples for each phase of the survey, develops sample weighting procedures, and provides assessments of the sample designs.
- The data collection branch has responsibility for computer-assisted survey information collection (CASIC), a common system for editing survey results. NASS partners with ERS on the tasks of questionnaire design and defining data edits. Specifications for data entry using CASIC and editing (within sample unit) are located in the data collection branch.
- The survey administration branch (SAB) provides administrative instruction and coordinates all data collection activities conducted in the NASS field offices. The survey administration branch has responsibility for questionnaire content through collaboration with ERS; and scheduling, training, edit testing, completion of interviews (questionnaires), and monitoring response rates. SAB provides project management oversight for coordinating the development of survey specification documents, coordination of project resources, and for preparing the OMB Docket requesting approval to conduct the survey.
- The statistical methods branch in the statistical division develops the cross-record edits to identify outliers and produces the aggregate survey indicators and estimates.
- The research branch develops the procedures for calibration and variance estimation and evaluates the effectiveness of these procedures for providing a measure of reliability of the aggregate estimates for each variable.

The ERS representatives on the steering committee focus on the agency's areas of responsibility: responding to congressional and USDA mandates, informing policy decisions, supporting national accounts, and providing information for USDA program management. In addition, ERS interest focuses on the impact of survey design and methods on the relationships between variables as assessed in econometric models.

The steering committee is the decision-making body for survey requirements and adding or subtracting questions, although issues can go to NASS and ERS management if the steering committee cannot come to an agreement. This formal mechanism is designed to assist the agencies in coordination and information sharing.

Larger decisions, such as designing a new sample for an extra survey or a new program, are made at the senior management level of NASS and ERS involving the administrators, since it may require extra funds to be allocated. However, there are limited opportunities for these major decisions to be made, since it is difficult to make changes once the survey is in the field. The steering committee has proven to be very protective of ARMS in managing its respondent burden and survey content.

MANAGING THE CHANGING FOCUS OF ARMS

These budgeting and survey management coordinating mechanisms have worked well by ensuring a common approach by the two agencies during the gestation and maturing periods of the survey. However, given the differences in perspective of NASS and ERS, the current management framework may not be sufficient to ensure that the survey will have the capacity to change to meet future data needs or to coordinate the increasingly complex technical and methodological environment that may define its future. In the future, the survey is expected to focus less on descriptive data and more on the type of high-level multivariate analysis represented by ERS and its constituency.

Other models of survey management employed for other major U.S. government surveys may have potential for better focusing responsibility for directing and funding ARMS for the future. One mechanism is the notion of survey sponsorship, in which the agency that is the primary consumer of the information from the survey has responsibility for its overall direction and, critically, for securing funding. A number of Census Bureau surveys, most notably the Current Population Survey (CPS), are managed with this sponsorship model.

The CPS is the monthly household labor force survey for the United States conducted by the U.S. Census Bureau for the Bureau of Labor Statistics (BLS). In the case of the CPS, BLS has primary responsibility for determining the labor force and other socioeconomic concepts and definitions, as well as the questions to be posed, in collaboration with experts at the Census Bureau. In order to fulfill this responsibility, BLS has assembled a small, highly skilled staff of mathematical statisticians and cognitive scientists to support its oversight role, again in coordination with the experts in these disciplines at the Census Bureau.

These resident skills permit BLS to communicate its requirements for information to support its program of research, analysis, and publication in a sophisticated manner. For example, data requirements are developed with specific coefficients of variation incorporated, and the trade-offs between cost (sample size) and reliability are discussed in those terms. Importantly, BLS has the responsibility for securing the bulk of the funding for the CPS, with funding for only the March income supplement and other specific collections used mainly by the Census Bureau remaining in the Census Bureau budget. Funds are transferred from BLS to the Census Bureau annually through the formal mechanism of a cooperative agreement, which establishes targets and standards for the survey.

Sponsorship arrangements have the benefit of aligning use with cost. In the context of the federal government, they cause the ultimate customer to justify the information sought through a formal budget defense process and to be responsive to other governmental processes that seek to ensure that performance measurement is associated with resource inputs.

In considering the current cooperative survey management structure for the ARMS program, the panel notes the potential opportunities of a sponsorship model that could clarify responsibility for the survey and more closely align the resources with its end uses. A potential sponsorship model for ARMS would include an increased role for ERS in developing the ARMS program, which may involve a shift in funding through the ERS budget (except perhaps for funding for NASS-specific products to meet NASS mandates). It would certainly require changes in staffing. For example, ERS may need to deepen its bench of mathematical statistics and cognitive science skills in order to assist in overall direction of the program.

The panel lacks expertise in organizational design to assess the benefits and costs of recommending a switch to a sponsorship model or any other particular model, such as the model employed by the National Center for Health Statistics and the Agency for Healthcare Research and Quality with regard to joint management of the Medical Expenditure Panel Survey (MEPS) and the National Health Interview Survey (NHIS), which provides the sampling frame for the MEPS. However, the shifting priorities for use of the data require a management structure that will ensure that the survey continues to be responsive while maintaining an appropriate emphasis on data quality. The Department of Agriculture may wish to investigate the sponsorship model as one way these shifts may be managed.

Stakeholder Feedback

Although coordination and input mechanisms in USDA are formalized and fairly sophisticated, the same cannot be said for its arrangements for obtaining, filtering, and implementing input from stakeholders and users. In many agencies across the federal government, the mechanism for facilitating communication with outside interest communities in an open and structured manner is the advisory board. However, the advisory board mechanism has not proven to be very effective in the case of ARMS.

The official NASS Advisory Committee on Agriculture Statistics is an organized and active body, but it is mainly focused on the Census of Agriculture and its related programs. This is understandable, since the NASS advisory committee was inherited from the Census Bureau when the Census of Agriculture was transferred from the U.S. Department of Commerce to USDA. The advisory committee does, from time to time, consider other issues. However, a review of all previous agendas and recommendations from the NASS advisory committee can find no reference to the advisory committee's reviewing ARMS content per se.

Nonetheless, ARMS is connected to the Census of Agriculture in the year of the census through coordination of the sample and questionnaire design, so the advisory committee has addressed issues of consistency between ARMS and the Census of Agriculture. The advisory committee did recommend in the 2002 annual meeting that NASS proceed with efforts to integrate concepts and processes of the agricultural census, ARMS, and related year-end surveys. This recommendation was implemented, with the result that operations selected for ARMS in 2002 were not mailed a census form. The census data were generated from ARMS. NASS has decided that a similar process will be used for the 2007 Census of Agriculture. Although this process reduces respondent burden, it does raise concern that if the questions or the context of questions differ in any respect between the two forms, then the estimates from the two questionnaires may well be different, resulting in bias in the census estimates.

In the absence of a formal advisory mechanism, both NASS and ERS conduct periodic but largely ad hoc meetings with data users outside USDA in an effort to receive feedback on their statistics and research programs and to hear what policy issues need to be addressed. These user forums are sometimes cosponsored and may be specific to different types of data, such as crop or livestock production, economic, or environmental data. The feedback from these meetings is channeled to the ARMS steering committee if it relates to issues that the ARMS program might address.

Over the years, USDA has solicited input from outside users in a variety of ways. One primary mechanism has been through informal cooperative arrangements with the American Agricultural Economics Association (AAEA). As a professional association of scientists who use economic tools to analyze issues and solve problems in the area of agriculture, food, and environmental resources, the AAEA has long maintained an Economic Statistics and Information Resources Committee (ESIRC), which takes a close and continuous interest in ARMS as well as the other products of NASS and ERS. This committee has at times taken on studies and fostered research and analysis designed to address difficult conceptual and measurement issues in agriculture.

A primary example of this informal collaboration is the *Commodity Costs and Returns Handbook*. This monograph was prepared by a task force organized by AAEA's economic statistics and information resources committee, on the basis of recommendations stemming from an AAEA conference in 1991. The mission given to the task force by the committee was "to recommend standardized practices for generating costs and returns estimates for agricultural commodities after a careful examination of the relevant economic theory and the merits of alternative methods." Most of the recommendations of this task force were subsequently adopted by ERS.[2]

The AAEA can provide a convenient base for collecting informal and current input on issues. For example, to obtain user input to this report, the panel sponsored a special user forum at the 2006 annual meetings of the AAEA. The feedback on the research uses of ARMS data, survey needs, and access requirements was invaluable in identifying issues to be addressed by the panel. ERS and NASS may want to jointly sponsor such forums on a regular basis in the future.

Another way to gain outside input on program content and priorities is through the ARMS online briefing room, which is maintained by ERS. This innovative briefing room provides a mechanism for user feedback and often broadcasts requests for input on specific aspects of ARMS.

Recommendation 3.1: The ARMS program should have structured mechanisms in place for stakeholder feedback and discussion on ARMS, beyond what is currently done, such as organized stakeholder forums, with some obligation to respond. Specifically, USDA should solicit input in developing the survey from stakeholders from within USDA and from other government agencies, universities, professional associations, and the private sector.

Recommendation 3.2: The NASS Advisory Committee on Agriculture Statistics should expand its scope to include an annual review of ARMS.

Research and Development

Every official statistics survey program that operates on a continual basis should have associated with it a methodology research and develop-

[2]Presentation by Bill McBride, Economic Research Service, January 18, 2007.

ment program. The jointly sponsored ARMS program suffers because such a program has not been formalized. This is not to say that there has not been some research and development during the decade of ARMS and its predecessor programs. There have, in fact, been some pieces of quite good research. However, these have been initiated primarily when a major change in the ARMS program has occurred or has been proposed. For example, when a mail economic questionnaire was introduced to increase the sample size, there was an effort to design a self-administered questionnaire to use alongside the interviewer-administered instrument. Similarly, a need arose for easier implementation of the complex sample design into calculations of variance to use in tests of significance, and so the delete-a-group jackknife procedure was developed.

An official statistics program should have an established time frame during which changes may be made to the sample and survey design, the questionnaire, the resulting editing and processing system, and the weighting, estimation, and other adjustment systems (such as statistical disclosure or seasonal adjustment). The survey design would be a factor in determining what an appropriate period of time is for changes to the design. Once this is set in place, a research and accompanying development program would be put in place to operate to the time schedule.

The ARMS program is in need of this formal structure to address both survey methodology and analysis of the results of the survey to inform policy-relevant issues. In view of the fact that ARMS is a jointly funded survey, the cycle for survey design changes and the methodology research and development program should be a jointly established and governed program. Joint goals should be set. The panel observes that many of the questions the agencies posed to us could have been answered by one or both of the agencies if such a program were in place.

As noted in Chapter 2, there has been a marked increase in interest in doing econometric research with the ARMS data. This has resulted in new requests for statistical support directed to an already-overtaxed NASS methods unit to facilitate analysis of complex survey data. These needs should be factored into the methodology research and development program to ensure that the data files and paradata are in place to support the appropriate statistical analyses for the complex survey data. More importantly, resources need to be made available *in both NASS and ERS* to support the program of statistical analysis of survey data, which is increasingly based on highly sophisticated techniques and methodologies.

Establishing joint governance of such a research and development program for ARMS would enhance the capabilities of both organizations to provide input and resources. As part of the methodology research and development program, both agencies might consider some options as to how they might work together in a more collaborative mode. It is conceivable that the two agencies might be able to fund one or more researchers through the American Statistical Association/National Science Foundation program to contribute to this program.

Such a joint program would ideally involve some skill transfer between the two agencies. Neither organization has many individuals with competence in the main skill set of the other agency. If NASS were to hire a few economic researchers familiar with survey research, and ERS were to hire a few mathematical statisticians with some background in economics, these individuals might help to provide a bridge between the survey focus of NASS and the econometric focus of ERS. An additional option (and possible interim solution) might be for the current senior mathematical statisticians at NASS and senior economic researchers at ERS to be sent to the sister agency to sit with agency staff for several days in each pay period.

Many models for the kind of research and development program envisioned here exist in the federal statistical community. For example, with respect to research and development on questionnaire and data collection processes, there are ongoing programs that are regularly revised on the basis of studies by in-house methodologists: the CPS and the Consumer Expenditure Survey at BLS; the decennial census, the American Time Use Survey and the Survey of Income and Program Participation at the Census Bureau; and the National Health Interview Survey at the National Center for Health Statistics are some examples. The research typically involves a mix of laboratory and field activities (including the analysis of paradata, which we discuss in Chapter 5) conducted by methods researchers, usually with advanced degrees in the social sciences. Some agencies rely heavily for graduate training of existing employees on the Joint Program in Survey Methodology (JPSM). Over the years, this program has strengthened the methodological programs in all of these agencies and would be a useful resource for NASS and ERS as well. JPSM primarily produces master's-level graduates and is expanding its Ph.D. program.

The kind of research program we are advocating requires at least some Ph.D.-level staff, because such training leads to greater expertise in conducting methodological research. Finally, staffing this effort in-house is important because internal researchers understand the problems in a way that contractors rarely can.

Recommendation 3.3: ERS and NASS should establish an ongoing, jointly sponsored, and appropriately funded methodology research and development program. The program should provide adequate resources to support current and future research, development, and statistical analysis needs throughout the implementation of ARMS and to assess and manage the quality of the data. If new funds cannot be obtained, funds from existing programs must be reallocated.

Need for a Long-Term Strategy and Plan

As with most government survey operations, as ARMS has matured, it has settled into a comfortable repetition of tried and true formats and collection rounds that have been periodically revised. Some significant strategic decisions have been put in place over the years, such as the closer alignment of ARMS with the 5-year Census of Agriculture. However, this has occurred largely without the discipline of a structured, long-term planning process.

In the larger picture, the managers of ARMS have other venues for enhancing the use of the data for econometric policy-relevant analyses. One such venue occurs periodically with the reauthorization of the farm bill. Although ARMS data, as they are, are very useful for evaluating policies implemented by this bill, the cycle of reauthorization affords a once-each-five-year opportunity for ARMS to reach out and preemptively to develop a plan for collecting farm bill–related data in a policy evaluation framework, with an emphasis on specific policies of interest. For instance, adapting a pre- and postsurvey component for ARMS in conjunction with policy implementation or working with administrators to evaluate a randomized trial of some farm programs could dramatically enhance the value of ARMS as a policy analysis tool.

For this and other reasons, it makes sense for ARMS to operate with a five-year plan in order to fit into the five-year cycle of the Census of Agriculture. Within each cycle, there is reason to hold the basic survey relatively constant, with changes permitted on an annual basis to add extra modules outside the core survey questionnaires or to enumerate follow-on surveys within the ARMS sampling frame. Thus, for example, inserting a question into the core set of questions should not be permitted within a given five-year cycle, except under extreme circumstances. More significant changes designed to maintain the relevance of ARMS in a changing farming environment could be made once each five years, with provision made for bridging the old and new times series in a manner that would enhance the value of the time series data to users. This would have a number of benefits, in addition to stabilizing the time series for data items of interest. For one thing, it would diminish the problems of recoding data items from year to year, which has had a confusing effect on users of ARMS microdata.

Recommendation 3.4: NASS and ERS should commit resources to developing a five-year plan tied to the Census of Agriculture for ARMS content, coverage, and methodology. The agencies should develop measures to control changes during the five-year period to minimize disruptions to the time series of the core content in ARMS.

4

Sample and Questionnaire Design

In this chapter, the panel considers issues of design and development of key methodological aspects of the Agricultural Resources Management Survey (ARMS), focusing on the sample frame for ARMS, the foundation and implementation of the multiphase stratified sample design of the survey program, and the design and development of the survey questionnaire. These methodological issues are addressed in light of survey goals and are evaluated in consideration of the panel's understanding of the current state of the art in survey and questionnaire design.

The panel recognizes that the survey has evolved over the years into an "as-is" state that must be taken into account when considering possible improvements. In summary, ARMS is executed in three distinct but interrelated phases. Each phase is complicated; together they form a complex mosaic of question variations, differing modes of administration, and tailored sample designs to represent the different populations covered. All of these factors must be considered jointly in the overall design process.

- The Phase I survey (screening survey) serves to screen a standing list of farms for commodities of interest, as well as for whether or not the farm is in business. The Phase I screening is sometimes combined with the screening for vegetable chemical use survey or the crop/stocks survey. It is commodity-specific and has up to 48 state-specific versions so as to avoid asking respondents about commodities not usually grown in their areas. It employs phone, web, and mail data collection modes.
- The Phase II survey (production practices survey) covers the use of chemicals, fertilizers, and pesticides and has from one to four versions of

the questionnaire, based on commodity. As noted earlier, the ARMS Phase II questionnaire was recently and temporarily integrated with another survey questionnaire for the Conservation Effects Assessment Project (CEAP) for those operations that were selected for both the ARMS and CEAP samples. It has only one collection mode: personal interviews via face-to-face contact.

- The Phase III survey (cost and returns survey) asks about farm and household economics and farm/farm operator characteristics, includes several questionnaire versions (a general questionnaire, from one to three commodity-specific questionnaires, and a core questionnaire), and includes an additional number of sample units added specifically to produce sufficient reliability to produce estimates for the 15-state oversample. The sample for this phase is drawn from both the list frame (also the source of sample units for Phases I and II), and an area frame using a sample selected from eligible farms identified in an annual June area survey. Every five years it is integrated with the Census of Agriculture. It incorporates several modes of data collection (face-to-face, telephone, mailout-mailback with face-to-face follow-up for the mail nonrespondents, and, soon to come, the Internet).

Each phase of the survey operations is a survey design challenge in itself. When combined in one program, they pose an intricate, interwoven series of design challenges that must be addressed holistically. It is useful to consider those challenges in light of sampling and questionnaire design goals, such as minimizing respondent burden, minimizing cost, and ensuring compatibility among the pieces and across time.

SURVEY DESIGN GOALS

Minimizing Respondent Burden

In a survey with so many phases and lengthy questionnaires on highly technical topics, the issue of respondent burden is pressing. The response burden (in minutes) for ARMS estimated by the National Agricultural Statistics Service (NASS) varies substantially for the different components of each survey phase (Table 4-1). The costs and returns survey (Phase III) is especially burdensome, containing questions that are difficult for the respondent to answer—often because the data are hard to retrieve or estimate and sometimes because the question is conceptually complex or unclear from the respondent's point of view.

These observations have several implications. One is that the survey suffers from both item nonresponse—that is, missing values of variables—and from unit nonresponse—that is, entirely missing observations in at least one phase. Another consequence is that there has been a conscious effort

TABLE 4-1 Estimated Response Burden by Survey and Phase, 2006

Survey	Minutes Per Response
Integrated screening survey (Phase I)	15
Vegetable chemical use	32
Practices and costs (Phase II)	57
Costs and returns (Phase III)	83
15-state core, costs and returns (Phase III)	57
Commodity costs of production (Phase III)	105
Organic soybeans practices and costs (Phase III)	57
Organic soybeans costs and returns (Phase III)	105

SOURCE: National Agricultural Statistics Service, submission to U.S. Office of Management and Budget, 2007.

on the part of survey managers to avoid repeat visits to the same farm in successive years when a farm is included in ARMS. The decision to avoid repeat visits limits the ability of the survey to follow farms longitudinally. As discussed later in this chapter, ARMS employs a special sampling routine (called Perry-Burt or P-B, see below) to reduce the likelihood that a farm will be selected for another survey in the survey year or the ARMS survey two years in a row. A further result is that the scope of the survey is circumscribed beyond what analysts consider desirable.[1]

Minimizing Costs

ARMS is a very expensive program. In an effort to hold down those expenses, as the survey has evolved, many steps have been taken to build in efficiencies and control costs. The reliance of the survey on many of the same respondents for more than one phase of data collection reflects, in part, an attempt to achieve collection efficiencies. The use of mixed modes for data collection is often an important means of controlling costs. The cooperative agreement arrangement with the National Association of State Departments of Agriculture provides a substantially more economical means of collecting data than would be possible working with any other data collection organization, in the view of NASS.[2] These and other measures that are designed in part to achieve cost efficiencies have potential implications for data quality.

[1]Presentation by Katherine Smith, Economic Research Service, February 2, 2006.

[2]Presentation by Robert Bass, Census and Surveys Division, National Agricultural Statistics Service, June 8, 2006.

Ensuring Compatibility

To ensure that the survey is comprehensible to the respondent and that there is theoretical coherence to the concepts employed by the analysts of the ultimate data, it is important that ARMS use a common conceptual framework across the phases and versions of the survey. This means that ARMS must establish consistent concepts and definitions among the phases and versions and that they should be coded for retrieval in a consistent manner as well. For the most part, ARMS has been successful in ensuring consistency of concepts and definitions, but there are exceptions. Since consistency of concepts and definitions relies on the common wording of questions and common formatting from version to version, the impact of truncating the number of questions in the self-administered or core questionnaire (version 5) in Phase III is of concern. For example, do the answers to fewer questions obtained by mail in the core questionnaire compare directly with the more detailed answers to the questions obtained by face-to-face interviews in the other versions of the Phase III questionnaire? These issues are discussed in this chapter.

Similarly, there are variations in compatibility across time. ARMS collects data on production practices for a rotating list of specific agricultural commodities, meaning, in practice, that some of the questions must change from one survey round to the next because production practices vary from commodity to commodity (see Table 2-1 for recent commodity coverage). Also, topics of special policy or research interest may be introduced and subsequently eliminated as the rationale for questions on the topics changes. These design features of the survey are sometimes incompatible with a desire to maintain a consistent core of questions over time. The goal should be consistency across time for commodities to the degree possible.

These changes in the questionnaires are likely to have affected the time series in unknowable ways. Particularly because the effects cannot now be quantified, they are of concern. Various design decisions could be made to increase consistency, such as implementing a panel design with consistent cost-of-production data on at least one commodity over a period of years to study dynamics over time, evaluate the effect of periodic changes in programs, and predict what factors change behavior.

SAMPLING FRAME

Target Population

The sampling frame is developed to ensure coverage of the population of interest (the target population) in the sample population. The target population for ARMS Phase III is the official U.S. Department of Agriculture

(USDA) farm population in the 48 contiguous states. This population includes all establishments (except institutional farms, see footnote 3) that sold or would or could normally have sold at least $1,000 (nominal) of agricultural products during the year.

This target population was originally established for the Census of Agriculture. By using this definition for both the census and the surveys, USDA appropriately ensures consistency between the census and the surveys. Furthermore, the definition of the target population has been consistently employed since 1974, so it has become ingrained as the appropriate population of interest. Nonetheless, a recent review of the Census of Agriculture concluded that the application of a target population that extends coverage to the very small farms with little overall effect on agricultural production imposes the significant challenges for NASS in finding its target population and getting that population to respond—and respond accurately—to the census and surveys (Council on Food, Agricultural and Resource Economics, 2007, p. 23).

Some of the changes in American agriculture outlined in Chapter 2 have increased the challenges imposed by this long-lasting definition of target population. The Census of Agriculture review likewise noted that the growth in large complex agricultural operations, integrated production, nontraditional farms and "life-style" farms have made practical interpretation of the definition a continuing challenge (Council on Food, Agricultural and Resource Economics, 2007, p. 13). While it is appropriate that these matters of definition be considered in the context of the Census of Agriculture to continue to ensure compatibility between the census and the surveys, the impact of the definition on ARMS should be recognized.

There are often exceptions to the general rule in statistical surveys, made for practical purposes. Some types of farming operations that might be considered to meet the farm population definition (e.g., "abnormal" farms) are not considered part of the ARMS target population. "Point farms," those with only the potential to sell more than $1,000 of agricultural products, are difficult to find as consistently as more clearly commercially oriented farms.[3] Such factors impose an added burden, since the exceptional types of farms

[3]A point farm is a farm that did not sell at least $1,000 of agricultural products during the year but could have. Point farms are included under the "would normally have sold" part of the farm definition. The determination of whether an agricultural establishment qualifies as a point farm is made by assigning specific point values for crop acreage and livestock inventory. Any establishment with at least 1,000 points will be defined as a point farm. Each assigned point is assumed to represent one dollar in potential sales. It is necessary to correctly identify these point farms to ensure their representation in the summary.

An abnormal farm is out of scope for the survey. It is defined as a business (i.e., operates land for agricultural purposes or with potential for agricultural production) that does not fit the criteria for the ARMS sample population. This includes Indian reservations, prison farms, private or university research farms, not-for-profit farms operated by religious organizations, and high school Future Farmers of America farms. These institutional farms do not have the same expenses or income patterns as traditional farms (National Agricultural Statistics Service, 2005, p. 4).

must be identified so that they can be contacted and asked scoping questions in order to determine if they will be included or excluded, as appropriate.

Dual Frame

NASS develops two sampling frames to select farms for ARMS and other periodic surveys. The primary sample is derived from the NASS list frame. The list frame for NASS surveys is different from and less comprehensive than the list for the agricultural census in that it does not contain potential farming and ranching operation that are available to NASS but have not yet been screened for agricultural activity.

Emphasis in constructing the list frame is placed on farms producing significant amounts of commodities for which NASS provides annual estimates of acreages, yields, and production. A special effort is made to identify and include cases in which a few holdings provide a large share of production of an important commodity, such as cattle in feedlots, hogs, poultry, potatoes, or rice. NASS attempts to keep the list frame as complete as possible, especially for the large producers. Recently, however, NASS has devoted extra attention to ensuring coverage of small farms and ranches and minority operators of farms and ranches. As recommended by a 1998 USDA National Commission on Small Farms, NASS has stepped up its outreach efforts into communities representing the small and minority farms as it constructs the list frame (U.S. Department of Agriculture, 2006a).

The list is constructed and maintained from many different sources, including other NASS surveys and administrative files, as well as third-party commercial databases and USDA program files. Names obtained from such sources that are not already on the NASS list frame are screened to determine their farm status prior to inclusion in the list. Records for approximately 1.3 million farms were carried on the list frame in 2005.

The second sampling frame for ARMS is the NASS area frame. This frame supports the NASS June area survey sample, which is constructed anew each spring. The ARMS selects a subsample of the June area survey sample that is not on the ARMS list sample and meets the official USDA definition of a farm. This process provides coverage of eligible farms that are not included in the list frame. The eligible farms not on the list frame are also known as nonoverlap farms (NOL).

In developing the area frame, NASS relies on satellite imagery, other aerial photographs, and maps to divide the U.S. land area into small segments. Each segment is about 1 square mile, and each has unique and identifiable boundaries. In most states, the segments are divided for sampling purposes into four broad land use categories classified by intensity of cultivation: land intensively cultivated for crops; land used primarily for livestock production; residential and business areas in urban areas; and areas devoted to parks, military installations, and other uses.

An initial area frame sample is randomly selected from these segments. The resulting probability sample contains about 15,400 area segments—roughly 0.7 percent of the total land area in the 48 contiguous states. Each June, NASS uses this area sample file to conduct a major multiple frame survey outside ARMS, in which about 52,000 farmers are visited by enumerators to get a firsthand accounting of their agricultural activities. This midyear survey identifies all land uses within the segment and collects information about crops, operator households, animals, grain storage facilities, and environmental factors. The resulting information can be used to stratify farm operations in the selected segments to target crops for follow-on surveys.

In principle, the area frame sample provides coverage of all agricultural activity in the United States, regardless of changes in farm boundaries and management. This sample frame construction technique tends to guard against omissions or duplication in the list frame. Indeed, in 2005 the area frame added about 1,600 eligible nonoverlap units to the frame for the Phase III ARMS when combined with the Phase I screened sample.

The dual frame approach used in ARMS has several benefits, as well as some possible drawbacks. There is no doubt that the area listing operation identifies a large number of small and other types of farms that are not identified in the more traditional listing operation. Many small farms are below the radar for local extension services and others who provide input to the lists. These farms often come into business and exit again relatively quickly. They share with other small businesses the characteristic of being hard to identify on a timely basis.

Given the farm population that is required to be covered by the sample, there are no meaningful alternatives to using the area frame for including omissions from the list frame. NASS does maintain a number of potential farm records on the list frame that are screened on a regular basis for agricultural activity. As mentioned previously, these potential farm records are included on the mail list for the Census of Agriculture but are not included on the sampling frame for ARMS. A relatively low percentage of these records are actual farms. Also, the number of potential farm records varies widely during the five-year census cycle, with the largest number only available in the year preceding the census. So, NASS has decided to only include records that have been identified as farms based on previous survey data as the sampling frame for ARMS, for consistency and efficiency reasons.

There is a question as to whether the benefits of finding these small farms are worth the cost. On one hand, inclusion of small farms, in the larger view, adds little to the estimates of the volume of production and the understanding of the overall impact of agriculture in the U.S. economy. On the other hand, there is a serious national concern about the well-being of small farms and farm families on those farms. Besides, the costs of developing the area frame for ARMS are marginal. The June survey, which bears most of the expense of developing the area frame sample, supports multiple survey operations and provides useful information in its own right. Still, if some way could be found to bring small farms and new operations into the regular listing operation for ARMS with more certainty, the now-marginal costs involved in the area frame operation could be further reduced or eliminated.

SAMPLE DESIGN

Three major objectives establish the sample design parameters for the ARMS: adequately representing all size classes, reducing respondent burden, and attaining an expected level of precision.

Sample Selection

The selection of the target population, described above, is largely done to ensure representation of all size classes of at least $1,000 of agricultural products. The target population defines the frame, which, in turn, plays an important role in defining the design of the survey. Translated into operational terms, ARMS covers all noninstitutional agricultural establishments with farm value sales (FVS) of at least $1,000 in agricultural products.[4] The design objective changes from year to year as different types of farms are targeted for the cost of production component of the survey. To accomplish these objectives, the sample frame is subsequently stratified by farm value sales and farm type (Table 4-2) for sampling and estimation purposes. The selection of the sample from the list frame follows these steps

1. The population is classified into five strata defined by farm value sales.
2. The Phase I sample is selected. The sample is selected independently by state. For each state, a systematic sample is selected within strata. The strata are formed based on farm value of sales. Within each stratum the population is sorted by type of farm before the stratified systematic sample is drawn.
3. The Phase I sample is reselected to eliminate duplication with other surveys. After the sample is drawn, poststrata are formed based on type of farm within the farm value of sales strata. The purpose of the poststrata is to control the way the Perry-Burt procedure (explained below) will reselect the sample to reduce respondent burden. If there are five FVS strata and 17 types of farms, then this will result in a maximum of 85 poststrata. The

---

[4]The farm value of sales is calculated by assigning points on a per-head/per-area basis that reflect expected sales.

TABLE 4-2 ARMS Sample by Farm Types and Sizes, 2005

Type of Farm	$1,000- $100,000	$100,000- $250,000	$250,000- $500,000	$500,000- $1,000,000	$1,000,000+	Total
Oilseeds, grains, beans	3,830	1,856	1,534	1,371	1,055	9,646
Tobacco	128	50	46	32	19	275
Cotton	65	131	188	269	304	957
Vegetables, melons & potatoes	172	101	131	187	746	1,337
Fruit, tree nuts & berries	795	358	281	291	563	2,288
Greenhouse, nursery	125	470	303	244	810	1,952
Cut Christmas trees	60	21	4	11	23	119
Other crops & hay	1684	241	106	98	131	2,260
Hogs & pigs	87	113	163	256	437	1,056
Milk	99	584	601	420	1,026	2,730
Cattle & calves	2,945	976	518	347	570	5,356
Sheep & goats	139	6	11	7	14	177
Equine	470	4				474
Poultry & eggs	85	347	708	1,203	1,228	3,571
Aquaculture	28	75	17	25	67	212
Other animal	73	19	8	4	8	112
Land/cropland						71
NOL (area)						1,610
Total	10,846	5,360	4,620	4,766	7,001	34,203

SOURCE: Economic Research Service.

Perry-Burt procedure cross-classifies these poststrata with the strata definitions for the other surveys as well as the strata definitions for last year's ARMS sample and reselects a sample within these cells that has less burden than the original sample. The poststrata ensure that only similar types of farms are replaced.

4. The Phase I sample is screened for target crops and "in-business" status. ("In-business" status means the screening survey response indicates the operation meets the ARMS definition for a farm operation [greater than $1,000 gross value sales or potential] for the survey reference year.)

5. The Phase III samples for costs and returns and the core versions are selected from the "good reports" to represent all agriculture in a state.

6. The Phase III list sample is then supplemented with farms that were found in the area list operation.

7. The Phase II commodity samples are selected from the "good reports" with target crops. ("Good reports" are screened samples that are in business and meet any other survey criteria such as farm type, commodity of interest, organic certification, etc., based on the commodity mix and questionnaire version for which the sample is targeted.)

After assignment into the FVS strata, the sample is further stratified by farm type. Each farm operation is classified into one of 17 farm types, and the type of farm forms the substrata within the design strata. The following types of farms were used for classification in 2005: oilseeds, grains, and beans; tobacco; cotton; vegetables, melons, and potatoes; fruit, tree nuts, and berries; greenhouse and nursery; cut Christmas trees; other crops and hay; hogs and pigs; milk; cattle and calves; sheep and goats; equine; poultry and eggs; aquaculture; other animals; and total land/cropland of all types. Table 4-2 shows the resulting ARMS sample for 2005 by type of farm and size.

Sample coverage varies significantly by farm type and size. The number of farms in the smallest size class ($1,000 to $100,000 FVS) was 10,846 in 2005, or less than 0.6 percent of the approximately 1.8 million farms in that size class. Large farms are oversampled. About one-fourth of the approximately 28,000 farms in the largest size class ($1,000,000 or more FVS) are in the sample. As might be expected, due to the thinness of the sample for many of the farm types and sizes, there is not coverage in each state.

Note that, left unadjusted, inflation over time will cause the nominally fixed dollar limits in ARMS to admit ever "smaller" farms, because the $1,000 limit of actual or potential income would fall in inflation-adjusted terms. Similarly, the stratification categories would change. Although these changes may be small in the short run, they are likely to have cumulated to a substantial change in a decade of operation of ARMS. To avoid this cumulative effect, it would be useful to fix the dollar amounts in the dollars of any period and henceforth hold them fixed until some event spurs more fundamental reconsideration of the sample design.

Strategy to Reduce Respondent Burden

The second major influence on the design is the objective of reducing respondent burden. As mentioned earlier, a major strategy for reducing burden is to avoid revisiting respondents from other NASS surveys and those who reported to ARMS in prior years.

NASS employs a method developed in the early 1990s by Charles Perry, Jameson Burt, and William Iwig to control sampling to minimize the number of times that NASS samples a farm operation for several surveys. Called the P-B method, it is designed to reduce the likelihood that a farm might be in the survey for two years in a row (Kott and Fetter, 1999). The P-B method cross-classifies the Phase I ARMS sample with the samples selected the previous year for ARMS and four other recurring USDA surveys—hogs, cattle, crops/stocks, and labor. The P-B method groups the four non-ARMS surveys to identify duplications across the surveys (first stage) and then groups the ARMS sample across years (second stage). As mentioned above in the discussion of the sample selection steps, the ARMS sample is then redrawn to have less overlap with the other surveys and with itself over years. Essentially, then, the Perry-Burt procedure cross-classifies these poststrata with the strata definitions for the other surveys as well as the strata definitions for prior years ARMS samples and reselects a sample within these cells that is less burdensome to the respondents than the original sample. The cross-classification is done within farm type substrata that are defined within the larger FVS strata. This is done to minimize bias in the final sample. Without the substrata, the P-B method could trade a one type of farm for another; for example, a nursery for a dairy.

Following application of the P-B method to selection of the Phase I sample, NASS draws the Phase II and Phase III list samples from "good" reports in Phase I. Beginning in 2005, NASS has used sequential interval poisson (SIP) sampling to select the samples for each phase of ARMS (Kott, 2003). As in Phase I, the objective is to reduce burden, so each operation is selected for one and only one sample. In the end, over 33,000 sample units in Phase III were from the list frame and about 1,600 from the area (NOL) frame.

There is some concern that the procedures used to reduce the probability of a respondent's inclusion in multiple surveys might lead to a biased sample. Is the set of cases that would otherwise be selected more than once systematically different from other cases? Does conditioning on selection under two sets of criteria say anything important about the cases, so that omitting such overlap might tend to induce systemic bias?

There has not been any investigation of potential bias induced by the P-B method since the initial analysis that led to the decision to employ it over a decade ago (Perry et al., 1994). The 1994 analysis concluded that the "potential for bias resulting from the second stage of the algorithm will be much less than one percent of the . . . estimates, hence undetectable in light of the coefficients of variation associated with the estimates." Since then, NASS has continued to assume that any bias would be overwhelmed by the size of the sampling error and would be undetectable.

NASS is considering several changes to ARMS sampling in the future. Research is under way to move to a multivariate probability proportionate to size (MPPS) design for Phase I. This design would allow more flexibility to further target the sample where it is needed most—rare and poorly represented farm types. The agency is also considering controlling burden and overlap with other surveys using SIP sampling. This would replace the Perry-Burt method of burden reduction (Kott, 2003). These important research areas are the type of work that is suggested for management under the interagency research and development program recommended in this report (Recommendation 3.3).

Level of Precision

A third major influence on the design is the specification of the expected level of precision of the key estimates. The expected level of precision defines the size and design of the survey sample (U.S. Office of Management and Budget, 2006a, p. 7). This level of precision is specified in target coefficients of variation—the ratio of the standard error for an estimate to the mean value of the estimate. A small coefficient of variation (say 1 percent) would indicate that an estimate could vary slightly due to sampling error, whereas large coefficients of variation would mean that the estimate is quite imprecise. The most common way to improve the coefficient of variation involves increasing sample size.

The expected level of precision for key ARMS estimates is set forth in NASS Policy and Standards Memoranda (PSM) 45 (Standards for Target Coefficients of Variation for Major Probability Surveys). Stratification of the eligible population, sample sizes, and sample allocations are determined to achieve the target coefficients of variation specified in this document, subject to budget constraints.

Table 4-3, taken from PSM 45, represents the NASS targets and applies to ARMS as well as other surveys. The PSM was first issued in 1999, before the program expanded to provide state-level estimates for the 15 states and updated in 2004. NASS reports that the agency meets 100 percent of the

TABLE 4-3 Target Coefficients of Variation for Expenditures from the Agricultural Resource Management Survey

Item	U.S.	Category*
Expenditures for:		
Total	2.5	8.0
Fuels, interest, farm services, seeds, taxes, fertilizers, chemicals	3.5	10.0
Feeds, labor, buildings and improvements, farm supplies	7.5	15.0
Livestock	10.0	20.0

*Maximum values for categories. A category is defined in three ways:
1. Region: Appalachian (KY, NC, TN, VA, WV); Corn Belt (IL, IN, IA, MO, OH); Delta (AR, LA, MS); Lake (MI, MN, WI); Mountain (AZ, CO, ID, MT, NV, NM, UT, WY); Northeast (CT, DE, ME, MD, MA, NH, NJ, NY, PA, RI, VT); Northern Plains (KS, NE, ND, SD); Pacific (CA, OR, WA); Southeast (AL, FL, GA, SC); Southern Plains (OK, TX).
2. Economic class: $1,000 to $9,999, $10,000 to $49,999, $50,000 to $99,999, $100,000 to $249,999, $250,000 to $499,999, $500,000 to $999,999, $1,000,000+.
3. Farm type: livestock, crop.

U.S. targets specified in PSM 45. As for the detailed targets, which include regional, each of the 15 states, and the expense classes, NASS met 86 percent of targets in the 2004 ARMS, 75 percent in the 2005 ARMS, and 97 percent in the 2006 ARMS.

QUESTIONNAIRE DESIGN AND DEVELOPMENT

The key linkage between concept and response is the design of the questionnaire. If done well, a questionnaire will yield information consistent with the concepts and definitions. If not, design of the questionnaire can be a major source of measurement error, defined as "the discrepancy between respondents' attributes and their survey responses" (Groves, 1987).

There are generally understood to be four sources of measurement error: the interviewer, the respondent, the questionnaire, and the mode of data collection and the data processing methods (Groves, 1987, p. S163-S166). In addressing issues of measurement error arising from the questionnaire itself, the panel undertakes to discuss best practices in organizing for and designing questionnaires.

In the past two decades, the science of questionnaire design has been refined to include two groups of specialists who are drawn on in a coordinated effort: content specialists and design specialists. In ARMS, the content specialists are mainly the subject-matter experts in the Economic Research Service (ERS) with program or survey development responsibilities, and the design specialists are mainly the NASS survey professionals who design and evaluate questionnaires, prepare training materials, and attend to the myriad of tasks pertaining to capturing information from respondents.

This is not a unique arrangement for a federal survey program. It is often the case that these specializations are based in different agencies, with each agency bringing its strength to the questionnaire design and evaluation process. When properly organized for the task, content and design specialists will perform as an integrated working group, constituting a questionnaire design and evaluation team that reaches out to incorporate the interviewer and the respondent in the process of selecting content and design through field and cognitive testing of each collection mode (Esposito, 2003).

In this section, the panel describes and critiques the questionnaire design process as it has been implemented in ARMS. Several concerns are noted with the current process, and recommendations are made for improvements. Although the panel has concerns about the highly technical nature of some parts of the ARMS questionnaires and about respondents' understanding of some specific questions, we view a detailed item-by-item review of the questionnaires as beyond the scope of this study.

Periodic Major Redesigns

Questionnaires used in all surveys conducted on a recurring basis need to be evaluated and revised from time to time. The need for revision arises because the topics of interest to the survey sponsors and data users may change over time, respondents' understanding of questions changes over time, the behaviors and opinions about which respondents might be questioned change over time, and so on. Without proper accommodation, all of these factors could reduce the statistical value and substantive relevance of a questionnaire. Some items may never have worked as intended, so periodic questionnaire revision provides an opportunity to improve them.

The process of changing questionnaire content is probably less daunting if there is a set of prescriptions or guidelines to follow. This reduces the revision task to one of matching existing items to the conditions in each guideline and then, when there is a match, taking the prescribed action. The problem with prescriptions and guidelines is that by definition they are general in nature and rarely fit particular questionnaire items in a straightforward way. Even when it is clear that they apply, the right action is not always clear. Sometimes a guideline may fit a questionnaire item in a superficial way, but because this approach does not take into account the constraints and nuances of the particular survey, the prescription may not actually be appropriate. Finally, the application of guidelines by themselves does not necessarily include an evaluation of their impact—have they helped or hurt or had no impact?

Annual Questionnaire Updating

NASS and ERS follow a more limited process for annual updating of the ARMS questionnaires. As discussed in Chapter 3, this is a shared responsibility between the two agencies. ERS provides questionnaire changes to NASS. The recommendations to add or delete questions are based on current policy issues and data requests from users. ERS writes a justification for each question. In some cases, questions that were tested but did not work in the past are subjected to further testing in this process. The NASS role is to determine the feasibility of the questions, considering question content, timing, space constraints, and do so in light of U.S. Office of Management and Budget (OMB) requirements.

Potential new or changed questions and combined questionnaires are subjected to at least a rudimentary cognitive review by NASS. The objective of the cognitive pretesting, mainly of paper questionnaires, is to determine if respondents are able to answer the questions, not to measure data quality. The agency utilizes two main types of cognitive testing: (a) observing enumerators interviewing respondents, with the observer probing for more information about apparent problems or general impression of questionnaire, and (b) using enumerators as test respondents. Following the fieldwork, the agency summarizes the results and makes appropriate changes.

This ongoing cognitive testing program is constrained by limited resources and by OMB rules requiring formal clearance of questionnaires involving 10 or more subjects. As a result, the cognitive testing is usually limited to fewer than 10 interviews.

NASS is considering several initiatives to strengthen the ongoing cognitive testing program.[5] One of these initiatives would be to obtain so-called generic OMB clearance for testing of questionnaires, following the example of several other agencies. OMB has approved such clearances for pretests, cognitive tests, and similar categories of information under which agencies are granted a continuing authorization to modify the instruments and information collected within the limits approved by OMB. Generic clearances require submission of applications for OMB approval and are processed in the same way as other clearances, but they provide greater flexibility for subsequent modification and a simplified process of notification to OMB of changes.

Other initiatives to improve the cognitive testing program include changing the emphasis to evaluating questions based on the quality of the reported data, not just ability to give an answer to the questions; conduct-

[5]Presentation by Kathy Ott, June 8, 2006.

ing multiple iterations of pretests; and using an independent contractor, as NASS did in the development of the mail version of the Phase III survey, to obtain an outside perspective.

While applauding these initiatives, the panel observes that other federal agencies follow a more extensive research and development approach to questionnaire design, with an ongoing, overall conceptual reevaluation followed by a theoretically guided redesign and rigorous empirical testing of the questionnaire. For example, the Census Bureau has codified standards for development and pretesting survey instruments and materials and has clearly demarked the responsibilities of program areas, the Statistical Research Division, and the Economic Statistics and Methods Processing Division. In a statement of policy, the Census Bureau comprehensively sets out standards and guidelines for all bureau programs (U.S. Census Bureau, 2003).

This deliberative process is often very labor-intensive, requiring highly developed skills in cognitive sciences and design. The development process is complex, often requiring a variety of methods and more than one iteration, as tests clarify both design issues and the conceptual framework that is feasible to probe. This more deliberative approach would better serve the ARMS program because it will produce more tailored solutions, the effectiveness of which is empirically grounded.

A recent effort (2003) to develop a short-form, self-administered mail version of ARMS Phase III for the 15-state oversample had some of the character of the methodology research and development-based approach the panel advocates. The objective was to redesign and improve the shortened ARMS form for mail survey administration, improving the design for self-administration and improving comprehension, incorporating instructions into the questionnaire, making it more user-friendly, and advancing the visual design. NASS engaged the Social and Economic Sciences Research Center of Washington State University for this work.[6]

In the redesign, the investigators considered the current state of science in understanding the linkage between how people perceive and respond to objects in their environment and desirable features of questionnaire design. They developed a series of visual design principles applied to the construction of this questionnaire.

The empirical test of the redesign, however, was not conclusive. The response rate, prior to enumerator follow-up, was only 28 percent, considerably lower than the final response rate achieved in the traditional modes of collection. In their search for a reason for the low response rate, the investigators wondered if the respondents' perception of length could

---

[6]Presentation by Danna L. Moore, Don A. Dillman, and Nikolay Ponomarev, June 8, 2006.

contribute to that lower mail response rate. Although the redesigned mail questionnaire was 16 pages, in contrast to the more than 30-page interviewer questionnaire, respondents may have been less aware of the survey length when the survey was administered by an enumerator. Moreover, the redesigned version is visually quite dense. These intriguing issues could not be answered in this limited redesign effort, but they could be considered in the context of an ongoing cognitive research and development program.

Although there was no experimental control group, NASS was able to conduct two data quality tests comparing the distribution of data for 15 variables and item nonresponse for 9 variables with the concurrently collected enumerator-administered version. However, these limited data quality tests did not permit any conclusion as to which questionnaire was better. Despite the inconclusive findings, NASS elected to continue to use the redesigned form for future surveys because "the redesigned form is much more visually appealing and user friendly as a self-administered instrument" (Ott and Crouse, 2005).

A continuing research and testing program could cognitively test respondents who complete questions to obtain their evaluations of the visual features of the questionnaire and the questions. Other research objectives could address mode effects, including the comparability of data collected through future versions of self-administered paper and web questionnaires with enumerator-mediated surveys.

Several issues warrant focused investigation in the continuing cognitive research and development program, regardless of interview mode:

1. What information goes into answers? What questions are answered by respondents on the basis of what they know versus what questions do they answer by searching their records and financial reports, and what questions do they answer by guessing? It is likely that both record checking and guessing might occur in a given interview. How do these different approaches affect data quality and what can be done to maximize quality for a given approach?

2. How well do ARMS concepts fit respondents' concepts? The ARMS questionnaires use fairly technical terminology that may be relatively unfamiliar to some respondents or may mean something different to them than to the question authors. These problems may be of particular concern for small farms. Can question wording be made less technical? What can be done—particularly in enumerator-mediated interviewing—to detect conceptual discrepancies and bring the thinking of respondents and question authors into alignment?

3. When do examples stimulate respondents' thinking by helping to define the concept and when do examples restrict respondents' thinking to just the examples?

UNDERSTANDING RECORD-KEEPING PRACTICES

In addition to understanding the effects of question wording and collection mode, it is important to understand the linkage among questions requiring reference to records, the existence of records, and records that are actually referenced. The kind of information NASS collects on all three phases of ARMS is based on hard facts about the farming operations and thus could be expected to rely heavily on records. Although some items may be quite familiar to every farm operator (acreage, management practices, hours worked, personal and family characteristics, and the like), it seems unlikely that farmers can accurately answer most of the inquiries on pesticide use and costs and returns without reference to some sort of written business record. Of course, the larger and more complex the operation, the more likely it is that responding to a question will require reference to a business or family record.

There is a general understanding of the role that records play in ARMS responses. At the conclusion of the Phase III survey, NASS has asked enumerators to record whether respondents looked at their records, how often they used those records, and which records were used. The concluding questions are tailored to the various questionnaires, since the form of records is understood to vary across operations. The costs and returns (Phase III) questionnaire, for example, asks what record was referred to when reporting most of the income and expense data: a general ledger or personal record book, a formal farm records book or account book, loose receipts, or a computer or computer printout.

In the case of some data items, the existence of records may be assumed. For example, farmers who apply pesticides are required by the Federal Pesticide Recordkeeping Program to maintain the necessary records of restricted-use pesticides to ensure the applicator's compliance with the regulation.[7]

Records for other important aspects of farmers' business may be less accessible. A study in Minnesota and Wisconsin found that one-third of farmers kept records only as needed for tax purposes, 43 percent used whole-farm record keeping, and just 2 percent did enterprise budgeting. The usual kinds of financial statements that are common in nonfarm businesses are often inadequate or missing altogether for farm businesses; an average of only 40 percent of all farm loan applicants were found to prepare financial statements, and 16 percent prepare business plans, according to lenders (Van Schaik, 2003).

One recent formal NASS study of respondent record-keeping practices gave equally discouraging results for a group of 96 farm operations in Missouri and Virginia that had previously reported on the ARMS Phase III survey. This 1998-1999 study, called the Panel Plus Pilot Study, arranged

[7]The 1990 Farm Bill mandated the secretary of agriculture to require certified private applicators to maintain records regarding the use of federally restricted-use pesticides.

for a supervisor to accompany the enumerator to the selected units to observe the interview and report back on operator reactions, availability of data, types of records, and the feasibility of using alternative data collection strategies (Ott, 1999).

The idea that some items could be answered without reference to records was affirmed in this survey: every respondent answered at least one core question using no records. About half used tax forms or loose receipts, about one-quarter used a computer, and less than one-fifth used a settlement sheet. The link between a solid source of data and the answer was quite tenuous. Almost 20 percent of respondents used no records at all for the entire interview. The conclusion of Panel Plus Pilot Study was that a relatively low percentage of farm operators have a formal record-keeping system.

In the absence of formal records, a significant number of respondents were observed to answer the questions by guesswork. Some questions seemed more susceptible to generating guesses than others—particularly expenses for utilities, farm labor hours, value of land and buildings, market value of equipment, household expenses for food, nonfarm transportation, and other living expenses. In some cases, respondents asked enumerators to help in making a guess, leading to the possibility of differential bias or variability across enumerators.

Although the few reviews of record-keeping practices that have been conducted have been relatively informal and based on very small samples, they are disquieting in that they cast doubt on the quality of the responses to several key data items in the ARMS. The potential impact on nonsampling bias of misreporting data to ARMS due to poor record-keeping practices may not be trivial.

Recommendation 4.1: The methodology research and development program the panel recommends should systematically (1) evaluate current instruments and practices, (2) collect data that inform both the revision of existing items as well as the creation of new items, (3) test revised instruments before they are put into production, (4) use experimental control groups to evaluate the differences between the old and new questionnaires, (5) improve understanding of respondent record-keeping practices and their effect on survey quality, and (6) designate a subsample of the existing ARMS sample for research and testing purposes. Key parts of this work would best be conducted in a cognitive or usability laboratory facility. It would be enabled by obtaining a generic clearance from the Office of Management and Budget for testing of all phases of the survey to allow for broader cognitive testing, evaluate the quality of data reported in response to each question, and evaluate the impact of mode of data collection across the three phases.

APPROPRIATENESS OF THE SURVEY QUESTIONS

Although the panel did not interpret its mandate to include review at the question level, we do think that certain questions warrant review in the context of the systematic methodology research and development activities we recommend. For example, the collection of income information in ARMS Phase III involves several questions about personal income.

As a general rule, personal income is among the most sensitive topics that are asked of survey respondents (Bradburn et al., 1989), and this variable seems to be unreported or misreported relatively often as a result (Moore et al., 1997). One technique that seems to help improve the quality of income reports is to allow the reporting of range values, rather than a single amount. For a respondent who truly does not have an exact answer, ranges may yield more honest answers. Experience with the Survey of Consumer Finances of the Federal Reserve Board shows that using ranges rather than point estimates virtually eliminated "don't know" as a response (Kennickell, 1997). Another such approach is the use of so-called unfolding brackets (e.g., Juster and Smith, 1997), in which respondents who are unable to or unwilling to answer an open income question are then asked if their income is above or below a particular and relatively low dollar figure (e.g., $5,000). If they say above, they are asked it is above or below a higher figure (e.g., $50,000) and, depending on their answer, they are asked if it is above another figure. The process continues until the respondent has assigned the income in question to a relatively narrow bracket.

Another approach is computerized self-administration, which is widely used to increase the reporting of sensitive behavioral information, such as drug use (e.g., Tourangeau and Smith, 1996). It is now common for interviewers who enter responses into a laptop computer in a face-to-face interview to allow the respondent to directly enter his or her answers into the laptop for the sections of the questionnaire that concern sensitive topics. In ARMS that could take place when collecting income data and possibly data on other topics. It is hard to separate the sense of privacy created by self-administration in general, whether paper or on computer, from the benefits of computerization in particular, for example, automatically selecting the next question depending on the previous answer(s) or flagging suspect or out-of-bounds answers. Computerization may increase respondents' sense of the study's legitimacy. This could increase a respondent's willingness to provide income data and facilitate response by automating such calculations as adding multiple sources of income. Thus, self-administered computerized data collection may improve the honesty, completeness, and precision of income data in ARMS.

Collection of income data is exactly the sort of methodology research and development issue that might be on the agenda of a dedicated staff of methodologists, as recommended above. Even if this kind of work is conducted and NASS and ERS determine that the accuracy of income data was not sufficiently improved by new methods to warrant a change, the decision would be empirically informed.

Consistency of Questions Across Survey Versions

ARMS Phase III is a multiversion survey. The commodity-specific versions (2-4) and the self-administered or core version (5) contain fewer questions than the main version (1), and the commodity-specific versions add questions concerning the commodity enterprise to the core version and are generally longer than version 1. Most of the questions are the same across versions. However, there are three classes of exceptions:

- When a particular "hot" topic is added in a given year (such as Internet use by farm businesses in 2005), only the main version includes these questions, not versions 2-5.
- Questions specific to the commodity enterprises in versions 2-4.
- In the section for farm debt, versions 2-5 contain fewer questions than the main version 1, but the common questions are not compatible across versions because questions on version 1 refer to specific loans, whereas questions on versions 2-5 refer to types of loans (in which a few loans may be combined together).

There are often good reasons for differing questions on different versions of the questionnaire. Asking fewer questions on ARMS versions 2-5 is probably done to reduce respondent burden. However, there are also trade-offs between collecting a larger sample (more respondents) and collecting information on more questions from the same respondents. These practices serve to limit research on particular topics to the subsample of version 1 respondents.

Consistency of Variables over Time

Questions in ARMS will need to change from time to time to meet unfulfilled needs, address new topics, and maintain relevance. When such changes occur, there is usually tension between the consistency of maintaining time series of variables and the flexibility of adding new questions. Clear procedures are needed to distinguish between core questions (which should remain constant) and noncore questions (which could change). Efforts should be made to keep consistent time series of variables. Table 4-4

TABLE 4-4 Variables Included in the Debt Section of the ARMS Phase III Questionnaire, 1996-2005

Year	Lender Type	Loan Balance	Interest Rate	Loan Term	Year Obtained	Percent for Farm Use	Loan Guarantee	Type of Loan	Purpose of Loan	Number of Other Loans	Balance on Other Loans
1996	✓	✓	✓	✓	✓	✓	✓	✓		✓	✓
1997	✓	✓	✓	✓	✓	✓	✓			✓	✓
1998	✓	✓	✓	✓	✓	✓	✓			✓	✓
1999	✓	✓	✓	✓	✓	✓	✓				
2000	✓	✓	✓	✓	✓	✓	✓			✓	✓
2001	✓	✓	✓	✓	✓	✓			✓	✓	✓
2002	✓	✓	✓	✓	✓	✓			✓	✓	✓
2003	✓	✓	✓		✓	✓	✓	✓	✓	✓	✓
2004	✓	✓	✓		✓	✓	✓	✓	✓	✓	✓
2005	✓	✓	✓	✓	✓	✓	✓	✓	✓	✓	✓

illustrates which questions have been asked consistently over the years and which have not.

Recommendation 4.2: ERS and NASS should improve the consistency of variables across ARMS versions and over time.

SUPPLEMENTAL DATA

As changes to the process of designing the questionnaire, NASS and ERS may wish to consider administrative arrangements and design changes that would enable a recurring supplementation of the microrecords of the survey with additional data items. One way to accomplish this would be to accelerate the program to enrich the data collected with data from other sources. Another would be to introduce a recurring, formal supplemental portion of the ongoing survey; such a supplement would require significant design changes.

The managers of ARMS have already done extensive work to enrich the microdata from ARMS with additional cross-tabulations and other insightful sources in order to respond to the call by researchers for additional data from the survey. For example, ERS already provides some additional data by adding a number of external variables to the farm business and household and crop production practices research microdata files. These variables include zip code, county and state codes, as well as administrative region designations. Other variables generated by combining ARMS data with other sources have also been added to the microdata records, such as cost-of-production estimates, farm typology, and commodity cost and return estimates.

In some cases there may be both a reduction in respondent burden and an increase in data quality from using administrative data to supplement or replace some survey data. Data on program participation, tax-related information, and geographically linked data obtained from other sources (such as satellite monitoring and local weather records) are obvious starting points.

In an attempt to use existing data and only ask additional questions that are needed, NASS draws on administrative sources for use in ARMS analysis and estimation. For example, NASS uses administrative data from the California Environmental Protection Agency's Mandatory Pesticide Use Reporting System and a similar system in Arizona instead of asking for such information from respondents. NASS reports that it is investigating the use of available USDA program payment data for potential use in its census and survey programs and is continually searching for new sources that would be helpful.

Even with these additions, much of what researchers need is simply not available on the questionnaire, nor is it available in a form that could be easily added to the individual record from other sources. Much of the needed data will have to be collected anew.

ARMS managers have put in place a capability to introduce ad hoc questions about "hot topics." A more formal recognition of the need to collect supplemental information as a regular part of ARMS might be useful. For example, a section of the questionnaire could be set aside for the collection of special items, and provision could be made for soliciting input from the general user community for items to be collected, perhaps on a cost-reimbursable basis. One model of such an arrangement is the Current Population Survey, which provides the opportunity to add questions with cost reimbursement by the organization that commissions the supplemental collection.

Because such data collection could be seen as ancillary to the central purpose of ARMS, efforts would need to be devoted to special training for enumerators and to motivating respondents about the particular importance of the data. The overall burden of collecting supplemental information might be reduced if collection were limited to specific subgroups, such as farms in a particular type of watershed.

Although ARMS is already perceived as a survey with a high level of respondent burden, additional data collection may well be justified if there are issues of sufficiently great importance that require joint analysis with other data already collected in the survey.

Recommendation 4.3: NASS and ERS should explore the collection of auxiliary information on a formal basis, as well as the feasibility of enriching the ARMS data files with information from administrative data sources, geospatial data, and the like.

5

Data Collection

The data collection process in any survey operation has a high impact on data quality. Along with the questionnaire and sample design, the data collection and capture processes can be major sources of measurement error—defined as the difference between the value of the variable provided by the respondent and the true, but unknown, value of that variable. Measurement error in the data collection phase of the survey may arise through distortions introduced via the mode of data collection, the effect of the interviewers[1] and their behavior on respondents' answers to questions, the effect of respondent interpretation of the questionnaire items, and the motivation of the respondent to provide high-quality answers (Federal Committee on Statistical Methodology, 2001).

In dealing with measurement error, as with other sources of error, statistical agencies are expected to design and administer their data collection methods in a manner that achieves the best balance between maximizing data quality and controlling measurement error, while minimizing respondent burden and cost (U.S. Office of Management and Budget, 2006a). In this chapter, several aspects of the survey operation that can contribute to maximizing data quality and controlling measurement error are addressed. We discuss the arrangements for collection of the data and assess the impacts of the various modes of collection, both current and potential. Finally, we discuss the need to capture and preserve information gleaned in the collection process and make recommendations for improving both

[1]In this section, the term "interviewer" is used instead of "enumerator" to better characterize the interaction during the interview session.

metadata (data about individual data items and questions) and paradata (data about the data collection process, whether from the respondent's or interviewer's perspective).

COLLECTION ARRANGEMENTS

The conditions for success in minimizing measurement error are established in the arrangements made for data collection and capture. The administrative arrangements should be documented, stable, and treated to continuous examination and improvement. The National Agricultural Statistics Service (NASS) has chosen to manage data collection by capitalizing on the strong foundation of a long-term relationship with cooperating state agriculture departments and, through that connection, securing the interviewer staff. The cooperative agreement the agency has with the National Association of State Departments of Agriculture (NASDA) has been the mechanism used for field data collection for the Agricultural Resource Management Survey (ARMS) and its predecessors, the Census of Agriculture, and all of its surveys since 1978 (National Association of State Departments of Agriculture, 2007). Prior to this cooperative agreement, federal employee interviewers were used for the surveys that preceded ARMS and other NASS surveys.

The NASDA mission is to represent the state departments of agriculture in the development, implementation, and communication of sound public policy and programs that support and promote the American agricultural industry, while protecting consumers and the environment. The cooperative agreement with NASS is one of three cooperative agreement programs that support that goal for NASDA.[2]

The cooperative agreement with NASS is big business for NASDA. In 2007, it was funded at approximately $27 million. In turn, the NASDA cooperative agreement is the largest cooperative agreement in the U.S. Department of Agriculture (USDA).

Under the guidance of the NASDA national office in Washington, 3,400 part-time (not more than 1,500 hours per year) interviewers are managed through a network of 46 NASDA state field offices—some of which represent multiple states. A total of 43 of the field offices have responsibility for conducting ARMS data collection (those in Alaska, Hawaii, and Puerto Rico do not).

The field staff of 1,400 interviewers and the office/telephone interviewer staff of about 2,100 are deployed and managed by about 520 NASDA supervisors, who are largely recruited from the interviewer pool. The NASS

[2]The other two cooperative agreements are with the Animal and Plant Health Inspection Service and the Foreign Agricultural Service.

role with the NASDA field operations reflects an arms-length relationship, much like other contractual government data collections. NASS provides guidance to the NASDA supervisory staff and has responsibility for training, but it does not have the authority to hire or otherwise discipline individual interviewers or supervisors.

This arrangement has not been opened to competition. When the panel questioned why NASS has not opened the data collection contract to bids from organizations other than NASDA, the response was that the arrangement is very cost-effective and that things have worked well to this point, so NASS is reluctant to change them. This question reportedly arises periodically and is addressed at the management level of USDA.

The state departments of agriculture themselves play a similar supporting role. State offices variously provide staff exclusive to NASS or other inputs (space, funds for printing costs, etc.) through the cooperative agreement. NASDA employees can also conduct telephone data collection, edit interviews, and transcribe paper questionnaires into the Blaise computer-assisted interviewing instruments from the state offices. The only difference in the role of federal and state employees working on NASS projects is where their paychecks originate. (Cooperative programs in NASS and other statistical agencies enable relationships that go beyond the usual hands-off one of a contractor with a government agency. The state employees enjoy a special relationship that extends to management of various survey functions.) There are about 1,100 NASS employees in headquarters and in colocation with state offices. The states have an additional 160 state employees devoted to statistical functions.

Among other things, the cooperative agreement states that the data provided by USDA/NASS are the official state data. By taking advantage of the economies of scale that NASS has in managing ARMS, the states are able to get higher quality data under the agreement than they would be able to assemble on their own.

QUALITY ASSURANCE

In this decentralized survey operation, NASS imposes quality measures and monitors the survey process to maintain the quality of the ultimate data from ARMS. The quality control methods used include recruiting and training, sample case control procedures, monitoring of interviewers, and data review. Monitoring is necessarily limited for in-person interviews, but there is a quality control plan to monitor telephone interviews on a sample basis.

NASDA supervisors, who report directly to the NASDA office in Washington, are a critical part of the administration of data collection and a key element in the quality control process. Under the cooperative agreement, supervisors hire and fire interviewers, and they review a certain sample of their work interviewers through recontacts. If a review of a case yields important inconsistencies or other data quality problems, it is supposed to be returned to the field.

Role of Interviewers and the Interview Process

The interviewer is a critical link in the quality chain. Several types of errors can occur in the dynamic setting of the interview through the interaction among the engagement and innate cognitive abilities of the respondent, the wording of questions, the diligence of the interviewer in following directions, and the tone of voice and personal mannerisms of interviewer. All are part of a complex interaction that characterizes the interview session, and all play a role in creating something commonly referred to as the "interviewer effect." In ARMS, for which many interviewers have substantial workloads, individual interviewers can have a large effect on the variance.

According to NASS, the interviewers recruited by NASDA supervisors are primarily from rural areas. They typically have an agricultural background and in fact often come from a farm household. The NASDA interviewers are a somewhat diverse group. Two-thirds of them are women. Although data on race and ethnic group were not available to the panel, we understand that blacks are not well represented. The gender and race/ethnic group of the interviewers is an important factor to consider when identifying the effect of the interviewer on the willingness of the respondents to participate in the survey and the reliability of their responses. Although experience indicates that gender does not seem to have a systematic effect on most types of collection of factual information, the evidence on the effects of the race and ethnic group of interviewers is less clear. NASS reports that it is trying to persuade NASDA to increase black as well as American Indian representation. This is a matter of some urgency for USDA, even though the full effect of interviewer race and ethnic group on response is not clearly understood. The potential effect of interviewers and survey methods on the quality of data on small and minority farms has been in the spotlight since issues raised in two landmark court case in the late 1990s (*Pigford v. Glickman* and *Brewington v. Glickman*) have heightened interest in the economic status of minority farmers.

Like all statistical agencies, NASS should be cognizant of the potential for interviewer effects. The agency should document interviewer assignments to individual interviews as a part of its normal data assembly, and it should use that information to deepen understanding of those effects and to develop means of controlling them.

Because many rural areas are relatively sparsely settled, it is not uncommon for interviewers and respondents to know each other. In many other government surveys, this situation would be systematically avoided. There are concerns that respondents may be inhibited about speaking forthrightly about personal information in front of someone known to them, or that interviewers may not strictly observe the survey protocol in such situations. In ARMS, only when interviewers feel uncomfortable interviewing acquaintances or respondents express discomfort with being interviewed is the issue taken to the supervisor and the case is reassigned. In some cases, the usual dynamics of acquaintance may inhibit respondents and interviewers from expressing their objection to the arrangements for the interview. However, NASS contends that it helps the response rate if the interviewer and respondent know one another. This is a topic that deserves further investigation.

Not enough is known about the interview effect on the quality of the data, especially in light of some of the unusual aspects of the interviewer-interviewee interaction in ARMS. A fuller understanding of the interviewer effect would require collection of additional data about the interview, as well as a scientific analysis of those data. This is the kind of methodology research topic that could be addressed by a dedicated staff of research social scientists.

Recommendation 5.1: ARMS should use automated means to collect paradata on interviewer assignments to cases, the relationship between the interviewer and the sample farm respondent (i.e., whether they know each other), demographic characteristics of the interviewer, and the characteristics of the sample farms for nonrespondents that are coordinated with information obtained for respondents, either through the interview or interviewer observation. These paradata could be used to determine the need for additional research on the impact of the relationship between the interviewer and the respondent on the quality of answers. This data collection can best be facilitated using computer-assisted technologies.

As for other surveys, training for ARMS interviewers covers techniques to gain respondent cooperation, questionnaire administration, and general record keeping. ARMS training also addresses conceptual issues, such as biosecurity, and interview skills, such as cultural awareness. Moreover, because the interviewers are given a chance to edit the paper questionnaires, they receive an unusual amount of training in the appropriate technical skills.

When using paper questionnaires, ARMS interviewers have flexibility in moving around the questionnaire (navigating) and varying the question presentation. It is evident from the feedback interviewers give and from observations that interviewers often move back and forth throughout the instrument and provide additional help to respondents in ways that are not usually allowed in conventional structured interviewing.

The assistance that interviewers provide extends to helping respondents with estimating a figure. To respond to many ARMS questions for which there is unlikely to be easily retrievable data at hand, the respondent will often need to engage in mental arithmetic or perform a paper-and-pencil calculation to provide an estimate. For example, responding to a question on the number of acres of soybeans may take some time to calculate, because many farmers have numerous fields of soybeans of varying acreages. Similarly, the conventional emphasis on adhering to the standardized question wording is not a top priority for interviewers. These deviations appear to be the result of several factors.

First, because many of the terms in ARMS questions are either technically complex or differ conceptually from the ways that some respondents think about them, interviewers are trained to provide definitions or explanations if a respondent seems unfamiliar with or confused by a particular financial or agricultural term. However, some respondents are given definitions and others are not, so the question stimulus differs among respondents, departing from a key requirement of orthodox standardized interviewing. Second, because interviewers are under pressure to complete interviews and respondents are sometimes impatient with the length of the ARMS questionnaire, the interviewers have an incentive to take actions that will tend to minimize the burden on the respondent. In particular, if an answer can be inferred from previous answers, the interviewers are likely to enter the response without even asking the question. Third, ARMS interviewers also depart from the script to improvise and ask additional questions to understand the respondent's situation. Such probing is often built into surveys, but it appears that much activity of this kind in ARMS is left to the initiative of the interviewer. Fourth, in the course of an interview, interviewers may learn that some items are best recalled by respondents in a different order than that specified by the questionnaire and some respondents may remember items or change their minds about earlier answers as a result of later explanations or the shift in context that can occur after exposure to additional questions.

These departures from standardized interviewing practices in ARMS may have exactly the intended consequence of ensuring that respondents understand the questions, interviewers understand the respondents' circumstances, interviews are as brief as possible, and the data are as accurate as possible. But then again they may not, and virtually nothing systematic is known about the effects of such practices in ARMS.

The potential behavior of interviewers in dealing with item nonresponse is also a concern. It is believed that interviewers sometimes work intensively with respondents to obtain answers. Such effort is encouraged, but interviewers are instructed not to fill in responses that the respondent cannot or will not answer. Nonetheless, there appears to be evidence that in some cases the interviewers assign values to what would otherwise be missing data, based on their information and their beliefs about how farms and farmers operate. There is a concern that such assignments are not documented. Moreover, it is entirely possible that interviewers might assign responses that are different from and inferior to what would have been assigned using systematic procedures used for other missing data. The current official procedure is that, when NASDA interviewers have any issues of missing data for which they think they have special knowledge, they should contact the state office for follow-up.

As indicated above, the interview process in ARMS is quite unorthodox for the federal statistical community, for which standardized, structured interviewing is the norm. For example, ARMS interviewers clarify survey concepts by providing definitions and examples when respondents seem to need this help, and they sometimes record answers in the order that respondents provide them, even when this is not the order in which the questions appear on the form. Standardized interviewing (e.g., Fowler and Mangione, 1990) rests on the assumption that all respondents are presented exactly the same questions (i.e., words) and all questions are presented in exactly the same order. The idea is that if the stimulus given to respondents is exactly the same, then differences between their answers can more easily be attributed to actual differences in the respondents' circumstances than if the question stimulus varies between respondents. Thus, providing definitions and examples, doing so with improvised wording, and providing this additional information to only some respondents conflict with the basic tenets of standardization. Similarly, recording answers that respondents provide "out of order," without reading (or rereading) all questions in the order they appear in the form, is also not standardized because the context (the immediately preceding questions) differs for different respondents.

This departure from strict standardization does not necessarily compromise data quality and may actually be appropriate for collecting ARMS data. But to our knowledge there is no evidence that directly bears on the impact of nonstandardized interviewing in ARMS. Stanley did study the interviewer-respondent interaction in the NASS quarterly agricultural survey and observed considerable departure from standardized interviewing, primarily to avoid violating conversational norms, such as being redundant (Stanley, 1996). For example, interviewers failed to read the introduction to questions 72 percent of the time that they should have been read, presumably because the design of the questionnaire called for the same introductions to be read identically numerous times. These interviewers also did not read the entire question 22 percent of the time, and they changed the wording of questions in a way that altered their meaning 19 percent of the time. Stanley argues that there were sound conversational reasons for doing this and that imposing stricter adherence to standardization might degrade the accuracy of responses. However, the study did not produce data that might bear on response accuracy.

There is a body of evidence collected by Conrad and Schober (Conrad et al., in press; Schober et al., 2004; Conrad and Schober, 2000; Schober and Conrad, 1997) that directly compares data quality (primarily response accuracy) in strictly standardized interviews to data quality in more flexible or "conversational" interviews. Under the latter technique, interviewers can choose their words to make sure respondents understand the questions as intended. Although these studies do not examine any items from ARMS (or agricultural surveys for that matter), the results may be instructive here. Across these studies, Conrad and Schober have found that allowing interviewers to clarify questions (primarily from federal surveys), using words of their choosing, improved respondents' comprehension and response accuracy, particularly when the circumstances on which they were reporting were ambiguous. For example, when respondents were asked how many people lived in a household that included one child away at college, they were far more accurate in those interviews in which interviewers could explain that a child living away at school is not counted in this survey; strictly standardized interviewers would not be able to provide this clarification unless they did so for all respondents, whether or not respondents asked for it. Conrad and Schober found no evidence that conversational interviews misled respondents or in other ways biased answers. The cost of this improved data quality was longer interviews due to the time required to provide clarification. The current approach to collecting data in ARMS has some of the character of Conrad and Schober's conversational interviewing and so may improve respondents' comprehension. However, we just don't know for sure.

For this reason, the ongoing research and evaluation program we recommend should systematically explore the ways that different interviewing practices may affect ARMS data. A related area of inquiry, which should be on the agenda of an ARMS research and evaluation program, concerns the origins of the information that respondents use when answering ARMS questions. While some information is currently collected about how respondents obtain this information, this could be studied much more systematically and in much greater depth. It seems likely that respondents simply know the answers to some questions, calculate the answers to others based on what they know (e.g., "Each shipment fills the back of my truck so the annual amount must be in the neighborhood of 60 tons"), and refer to records and financial reports for others. All of these approaches may contribute different types of error in the measurement process and deserve investigation. Review of reporting practices and the interaction between the interviewer and the respondent should be a part of an overall evaluation of the interviewing techniques used in ARMS.

As discussed in Chapter 4, respondents are expected to consult various documents, such as tax returns and USDA program documents, during the interview. Reference to documents is typically taken as a positive sign for data quality when there is conceptual agreement between the documents and the questions asked in the interview. More troubling is the report that interviewers and respondents frequently need to make calculations during the interview in order to provide an answer to some questions. Such actions tend to increase respondent burden, and, unless interviewers or respondents are experts in accounting, there may be a serious compromise in the quality of the data recorded.

Recommendation 5.2: NASS should systematically explore the consequences of interviewer departures from standardization in the interview. To facilitate this, NASS should collect paradata on the frequency with which interviewers follow the order of the questionnaire, read questions as worded, provide clarification, and similar indications of departures from standardized procedures.

Role of the Reinterview in Data Accuracy

The panel considers the control and measurement of data accuracy to be major issues for the ARMS data, especially for the cost-of-production and farm income figures. Some control and measurement methods are employed, and others, found useful in other settings and in prior incarnations of ARMS, are not.

As mentioned above, under the cooperative agreement, supervisors recontact a sample of respondents in the process of conducting a case review. NASS reports that the quality control recontacts are randomly selected for each interviewer and supervisory interviewer. Additional recontacts are made if problems are suspected or uncovered. Quality control recontacts are made by the field office survey statistician and by supervisory enumerators.

At the conclusion of the survey operation, as part of post-survey activities, the state statistical offices attempt to capture conceptual and reporting issues in the formal Survey Evaluations (Form E-2) that are forwarded to the national office of NASS. While important to ensure interviewer quality and provide insights on reporting problems, the current quality control procedure is no substitute for a program of systematic validation (reinterview) studies. There have been no reinterview studies conducted for ARMS in recent years.

A formal reinterview study is an important method for estimating and reducing nonsampling errors in surveys (Biemer and Forsman, 1992). It can, like the current recontact program in ARMS, evaluate the fieldwork by detecting and discouraging cheating and interviewer errors. It can also play a more comprehensive role in ferreting out nonsampling error by identifying content errors that should be reflected in a model of survey errors by estimating (a) simple response variance, or the variability in the survey estimate over conceptual repetitions of the survey, and (b) response bias, or the response errors that would be consistent over repetitions of the survey. Content errors include definitional problems, misinterpretation of questions and survey concepts, and reporting errors.

NASS has a rich history as a pioneer in implementing recurring reinterview studies in the agricultural surveys (Hanuschak et al., 1991). As early as 1975, formal reinterview studies were conducted to probe how well reporting unit concepts were understood by the respondent. (Probing questions in this early study found that approximately 30 percent of respondents incorrectly reported total acres operated, and 20 to 30 percent incorrectly reported specific livestock inventories [Bosecker and Kelly, 1975].) From the late 1980s to the early 1990s, several important reinterview studies were conducted to measure response bias in surveys that included quarterly grain stocks, the June agricultural survey, on-farm grain stocks, hogs, and cattle on feed. These surveys not only identified conceptual difficulties that could be remedied by changes in questions or in interviewer instructions and training, but also had the more practical application of informing the board estimation process with measures of response bias. Although expensive in terms of statistician and interviewer time and additional burden on the part of respondents, these formal reinterviews were considered a success and formed the basis for a panoply of recommendations for a more robust reinterview program in the future.

Particularly because of the complex nature of the ARMS questions and the unorthodox approach taken to interviewing, the panel is concerned about the lack of systemic and continuous collection of information about the response variance and response bias in the survey. If it is judged that the resource cost and response burden of an ongoing, formal reinterview study of a sample of respondents are too large, there may be other techniques that will yield useful estimates of nonsampling error in the survey. One simple technique would be to do recordings of the recontacts to be sure the interviews are structured. An alternative would be to conduct dependent interviews (i.e., providing the respondent his or her answers from the previous interview and checking for changes) rather than full reinterviews.

Recommendation 5.3: NASS should use available analytic tools, for example, cognitive interviews, interviewer debriefing, recording and coding of interviews, and reinterviews, to investigate the quality of survey responses.

COLLECTION METHODOLOGY

The ARMS program has not fully exploited the available technologies for data collection. Considerable progress has been made in enhancing the back office operations involved with data capture, editing, and processing, and there have been far-sighted projects to test data collection through the use of personal digital assistants (PDAs) and global positioning system (GPS) devices. However, the development of and conversion to an automated field data collection mode lags behind the state of technology in data collection.

One collection mode that stands out as especially promising for further development is computer-assisted interviewing (CAI). There may be institutional issues that explain the lack of progress by NASS here. USDA has held that there is no technology currently available that can efficiently collect data on farm chemical use, production practices, cost-of-production information, and detailed cost and income statistics (National Agricultural Statistics Service, 2005).

A factor that complicates technological advances in ARMS is the wide variety of collection modes that are currently in play. In Phase I data are now collected in 4 modes—primarily through telephone interviewing (with about a 75 percent telephone response), with the remainder split among mail response, experimental web collection, and personal interview. (Mail response has been low enough in some states that ARMS does not attempt to use this method there.)

ARMS has recently begun preparations for expanding reporting via the Internet. It is expected that a web-based instrument will be offered for the 2007 ARMS Phase I screening. One concern about a shift to this mode of data collection is that it may also induce changes in the ways that respondents answer questions. This planned change reinforces the importance of research to understand how respondents answer questions in the ARMS interviews.

There are currently no plans to develop a web-based instrument for ARMS Phase II or for the fruit and nut, vegetable, or postharvest chemical use surveys, since much of the data collected requires the identification of a specific farm field that is planted to a specific commodity, and this field identification reportedly cannot easily be made on the Internet. Also, the detailed chemical application data are often copied from farm records by the interviewer during the interview. Plans are that Phase II will continue to be collected solely through personal interviews. However, other computer-based technologies such as PDAs and GPS devices might be relevant for this phase. These technologies are discussed later in this chapter.

Research will commence on the development of a mail instrument for the full ARMS economic version (ARMS Phase III cost and returns), which to date has been collected through face-to-face interviews. This research will build on the already existing mail instrument that currently covers the core (version 5) Phase III questionnaire, and it is in anticipation of the coordination of data collection with the 2007 Census of Agriculture to be conducted in early 2008 (National Agricultural Statistics Service, 2005). In that regard, development is currently under way for a web-based version of the ARMS economic phase (Phase III) core questionnaire, which was first mailed to respondents in 2004. However, as with the proposed change to web-based data collection for Phase I, it is important that ARMS have in place a research program to identify and control any adverse effects of this change in interview mode.

Computer-Assisted Personal Interviewing

Until these research and development projects bear fruit, it is expected that the vast majority of the data in Phases II and III will continue to be collected through face-to-face interviews. That being the case, there should be serious consideration of automating the face-to-face interview process by using computer-assisted personal interviewing (CAPI). The basic idea of CAPI is quite simple—instead of having interviewers navigate though paper-and-pencil entry on their own, with CAPI a computer controls the logical flow of the interviews, presents appropriate versions of the questions to be read, and offers a place for the direct recording of the answers to the questions.

The possible use of electronic data collection for the agency's personal interview data collection surveys has been tested, evaluated, and discussed at length since the first CAPI test with the 1989 September agricultural survey (Eklund, 1993).[3] This first experiment was followed with a test of collection of the farm cost and returns survey in February 1991. The conclusions from these studies were very favorable to the adoption of CAPI:

- Interviewers can learn and use CAPI effectively, even for the most difficult surveys.
- The data quality showed potential improvement, particularly by ensuring that interviewers answer questions and enter them into the proper cells.
- Respondent reaction was mostly indifferent, but more positive than negative.
- Interviewer reaction was often initially apprehensive but turned enthusiastic as training commenced. The positive reaction, however, may not hold for all interviewers.
- CAPI costs compared favorably with the paper-and-pencil method.

[3]Another early research report on experimenting with CAPI in its monthly livestock prices received and June area surveys is Eklund (1991).

Now, over a decade after those pioneering tests of CAPI, NASS has not yet developed a formal business case analysis for the use of CAPI with ARMS. The agency acknowledges the potential of CAPI but does not expect to move to that mode of interviewing until 2009 at the earliest.

Cost is always an inhibiting factor in adopting new technologies in statistical agencies and, indeed, the primary constraint in moving to CAPI is cost, according to NASS. One major element of cost is the purchase of necessary data entry equipment. Although the inflation-adjusted price of laptop computers continues to decline, purchase of sufficient machines for the entire ARMS field staff would still require a substantial sum of money. Particularly in light of the relatively short life span of a computer, purchase could be justified only if the machines could be used for a number of surveys over the effective life of the machine. NASS reported that their principal surveys that use personal interviewing take place only once a year and that such surveys are few enough in number to render CAPI uneconomical. In addition, they cited pressure to reduce personal interviewing even more, in favor of increased telephone and self-administered mail and web electronic data collection. It is not clear how such arguments apply to leased computers. Some other organizations routinely use leased laptop computers to support CAPI.

With CAPI there is no data entry operation after collection and typically no opportunity for "data grooming" (i.e., manually revising handwritten information after the interview) by the interviewer. Thus, two time-consuming procedures could be eliminated. Straightforward implementation of the ARMS questionnaire in CAPI might not be faster than the current paper-and-pencil approach. Indeed, it is possible that it could be slower, if interviewers were still obliged to navigate back and forth through the instrument in order to record information appropriately.

Although the loss of the possibility of post-interview data review could be costly in terms of data quality, the logic-based structure of CAPI makes it possible to introduce systematic quality control checks during the course of the interview and to resolve them while the respondent is present. A debriefing interview with the interviewer to be filled out for each completed case, as is done with the Survey of Consumer Finances, could provide an opportunity for critical comments that interviewers were unable to record during the course of the interview.

For obtaining panel data, as the Economic Research Service (ERS) has indicated it would like to accomplish in the ARDIS initiative (see Chapter 2), dependent interviewing is a good way to detect and eliminate spurious data changes. With CAPI, such comparisons are straightforward. However, it should be noted that such an approach can produce mixed results, as in the Current Population Survey experience. When the CPS converted from paper to computerized data collection, dependent interviewing was introduced. This drastically reduced reported change in occupation from what were clearly spuriously high levels (Polivka and Rothgeb, 1993). However, the extremely low levels of change after dependent interviewing was introduced may lie below true levels of change, reflecting respondents' recognition that when asked if a change has occurred, reporting no change will lead to the shortest interview because there will be no follow-up questions about the new job.

Computerization of data collection makes it possible to do things that are either very difficult or impossible with a paper-and-pencil interview. The presentation of multimedia information is a straightforward matter with CAPI; for example, one might present the respondent with images of crops, pesticides, aerial photographs of fields, etc. When the Survey of Consumer Finances moved from paper to the CAPI system, the "don't know" responses for dollar amounts were virtually eliminated because interviewers were told to automatically probe the respondents according to a prescribed protocol (as noted earlier, the decline in "don't know" responses was accompanied by no significant change in the proportion of refused answers). Computerization also makes it far easier to collect paradata in a form that facilitates its use for methodological and substantive research, as we discuss later in the chapter. This would allow, among other things, the identification of items that are difficult for respondents to answer. Similarly, items that are given inadequate thought can be identified because their response times are too brief.

In very complex interviews, programming costs for CAPI may be substantial and considerable time may be required to debug the instrument. In a repeated survey in which revisions are gradual, costs are far lower in waves after the first one. Effectively, the cost of the initial programming would be amortized over many survey administrations. Similarly, other costs of transition, such as the programming necessary to extract the data in a form that would be comparable to existing processing systems, would largely be borne once.

In light of the now extensive experience with CAPI in other surveys, concerns about respondents' reactions to an interviewer who is entering information into a computer strike us as overblown. Farmers, like most people in American society, have come to accept—whether grudgingly or with open arms—the ubiquity of computers in everyday interactions. If there is reason to believe that farmers are a special case—and we do not think they are—then this warrants special study. We have no reason to think that introducing a computer into the interview would deter participation in 2007 or beyond.

The one place in which the move to CAPI in the 1990s clearly reduced cost was eliminating back-end tasks like data entry (e.g., Baker et al., 1995). Otherwise the cost was largely a wash. With ARMS, CAPI would eliminate both the data entry operation and data grooming by interviewer, both of which may increase salary costs due to increased time. Moreover, the reduced price and increased power of laptop computing since the early days of CAPI would almost surely reduce the cost of this transition in ARMS compared with the cost in the early days of the technology.

Web-Based Data Collection

Data collection via the Internet offers potentially large cost savings. Beyond the initial cost of programming the instrument, the marginal costs should be minimal. If such an approach is successful, it would be possible to increase the sample size substantially with minimal cost consequences.

However, any change of mode requires careful thought. The experience of ARMS with self-administered interviews is largely concentrated in the short version of the Phase III interview. A systematic investigation of possible mode effects in that questionnaire version should be a high priority, and it should certainly take place before considering more intensive web-based data collection. There may also be adverse perceptions of the privacy of data entered via the Internet that should be studied.

As with similar concerns about CAPI, the concern that farmers will not use the Internet because they lack the computer sophistication to do so also strike us as unfounded and demeaning. The Internet is part of modern life. In fact, it may be more important in rural than suburban and urban regions because it connects people with the rest of the world. Although at present, the farm population is somewhat less likely to have access to a broadband connections, the difference is apt to shrink in the near future.

Integration of CAPI and Web-Based Collection

Can developing a CAPI questionnaire reduce the costs of developing a web-based questionnaire? Because CAPI involves an interviewer but web-based collection is self-administered, it is not easy to directly use the CAPI instrument on the Internet without some modification. But in survey programming languages in which questionnaire "routing" (a specification of the logical path between objects in the interview) is created separately from question text, as is the case with Blaise and MR interview, it is likely that there would be substantial savings in web-based development costs and time by adapting a CAPI questionnaire. With careful planning, a simple change in the display format, which may be accomplished through templates and stylesheets, may be sufficient in many cases to render such an adaptation relatively simple. Even with less sophisticated computer languages, the question logic of both CAPI and web-based questionnaires is likely to be similar, as are the user interface decisions (e.g., check boxes for check-all-that-apply questions, radio buttons for mutually exclusive response options). Moreover, there is much overlap between the set of skills required to do both kinds of programming, so the same programmers can almost certainly do both.

As noted earlier, the ARMS data collection process as currently structured appears to require considerable flexibility by the interviewer—for example, the ability to navigate between questions in unanticipated order. If such nonlinearity cannot be eliminated or substantially reduced by additional questionnaire research, CAPI implementation might not be able to depend on a standardized approach to instrument development. Nonetheless, such systems have been developed and utilized successfully in other survey programs for instrument testing, and with creativity they could be applied to develop and electronically index to facilitate data collection in the field. It is likely that NASS would have to hire or train a small staff of dedicated interview programmers or to hire a firm skilled in electronic questionnaire development. Despite the possibility of difficulties and additional effort to computerize ARMS, we strongly think computerization is worthwhile.

Recommendation 5.4: NASS should move to computer-assisted interview and possibly web-based data collection, after research and testing to determine possible effects of the collection mode on the data. CAPI and web-based data collection will provide opportunities to increase timeliness, improve data quality, reduce cost, and obtain important paradata.

Electronic Devices in Data Collection

NASS has recently experimented with using electronic devices in personal interviews, such as for locating sample points with GPS devices in Washington State and collecting cotton yield objective survey data in North Carolina. The resulting research reports, which can be obtained from the NASS website, are summarized below (National Agricultural Statistics Service, 2007b).

In 2004, the NASS Washington field office and the Research and Development Division combined efforts to study the practicality of using handheld GPS receivers to augment the ARMS Phase II survey data (Gerling, 2005). Washington field interviewers were supplied with Garmin GPS-72 receivers to obtain latitude and longitude coordinates of the sampled fields, rather than using county highway maps and the DLG Map software, as had been done previously.

In general, the field interviewers had few problems using the GPS receivers. Of the 211 positive usable reports, 22 (10.4 percent) operations had fields that could not be accessed by the field interviewer because the operator refused permission to approach the fields or the weather conditions made the fields inaccessible. These fields were recorded on county maps and later transferred to the DLG Map software to obtain the latitude and longitude coordinates. On average, interviewers spent 20 minutes and drove 9 miles to obtain the coordinates of each field with the GPS receivers.

The use of GPS devices in data collection has particular promise for modernizing and enriching the Phase II data collection, in which a specific field is the sample unit. Locating the centroid of the field with a GPS device would add considerably to the data value of the information. This could be done at a surprisingly reasonable cost. The study estimated total data collection cost to implement GPS receivers for all states for the 2006 ARMS Phase II sample is at $127,264. The first year's annual cost would be in the range of $50,000, with subsequent years' costs affected by inflation.

PDAs are another promising technology. In another experiment, the North Carolina field office used PDAs to collect data for the 2004 cotton objective yield survey Form B data (Neas et al., 2006). The office developed a user-friendly data collection instrument to collect Form B data onto a PDA and securely transmit them. Results showed that field interviewers could successfully collect and transmit the data via a PDA, providing them more quickly, eliminating mailing costs, and improving the overall quality of the data, since edit checks were incorporated into the data collection instrument. However, conducting a cost-benefit analysis showed that use of PDAs in more field data collections and administrative activities would be needed to consider them cost-effective.

Finally, we note that a GPS device can be integrated into a PDA. This would facilitate collecting geographic information in the context of the PDA-driven data collection. This variant, wedding the promise of GPS for the Phase II collection with the advantages of portability offered by the PDA, should be tested as well.

Electronic Data Interchange

Electronic data interchange (EDI) would allow the direct uploading of a farming operation's financial records to a USDA database. This approach should be seriously explored as an alternative to conventional modes of self-reporting for Phase III data collection. This will require ethnographic research to understand the current practices of farmers so that a system can be designed to match respondents' record-keeping practices.

The record-keeping practice surveys discussed in Chapter 4 can provide information on the extent to which respondents maintain their records electronically. EDI may be an attractive option for some respondents, particularly those who would rather not sit through a lengthy interview. If so, this could increase response rates for this subset of respondents and reduce interviewing costs.

The experience of the Current Economic Statistics program at the Bureau of Labor Statistics, which includes substantial research and evaluation, provides an excellent starting point. There are, of course, many differences between these data (which requests just a few numerical entries) and ARMS data (which requests many items in many formats).

There is a growing usage of standardized electronic book-keeping and report preparation packages in the farming sector, and, as standardized electronic book-keeping and reporting systems are further promulgated, EDI has the potential of seamlessly collecting some common data items, often much more quickly and potentially more frequently than is now possible. There are, however, several potential roadblocks. Any effort to proceed with EDI would need to be sensitive to respondents' feeling that providing actual records directly to a government agency may compromise their privacy more than reporting to an interviewer on a question-by-question basis, in which the respondent and the interviewer are in control of the flow of information. Some respondents may require the continuing assurances or persuasion of an interviewer to maintain their cooperation. For those who want to use a more efficient way of sharing their data, EDI ought to be an option.

DATA CAPTURE, EDITING, AND PROCESSING

ARMS employs a multilayered process of data capture, editing, and processing. Interviewers perform an initial review of their interviews with the goal of correcting errors; a systematic review of the data occurs in the field offices; keyed data at data entry points is carefully monitored; NASS data review happens simultaneously with the field office review; and an outlier board with representation from both NASS and ERS reviews outliers.

Supporting this multilayered system are automated tools, both off-the-shelf and internally developed. PRISM, an interactive data review system developed by NASS, allows for interactive review of error listings from computerized batch edits and previously submitted data corrections. NASS also uses the Feith system to review scanned images of keyed questionnaires and the NASS-developed IDAS tool to review data at both micro and macro levels.

These procedures appear to be fully in keeping with standard practices for data capture, editing, and processing, and the high degree of sophisticated process automation appears to insure against generation of errors in these processes. A defect of the process is that information about changes to the data is not systematically retained or is not retained in a way that can support methodological research.

METADATA AND PARADATA

We have identified several points in the process of data collection in which errors could be generated—points that, at a minimum, should be transparent for full understanding of the meaning of the data. We have suggested that one way that NASS and ERS agencies can further assess quality and assist data users to evaluate the quality of survey information is through capturing and providing supplementary information, known as metadata and paradata.

Survey ***metadata*** provide context that can help in the interpretation and use of individual data items and statistical aggregations of them. The most basic form of such information is the text of the questions, including response options, that elicited the data, and any system of codes needed to understand the meaning of open responses. Among many other types of information, the question text provides the time period for which the respondent has reported activity. This can be critical in making sense of answers.

Another class of metadata consists of indicators that reflect the quality of the individual pieces of information collected, such as information on the original status of each item (whether it was reported fully by the respondent or was missing in a particular way), actions taken that altered the original item in terms of content or position in a set of data as a result of editing or any form of data processing, comparisons of values to parallel values from other sources, particular evidence from initial question testing and design that may bear on the content and reliability of the questions asked, among others. If answers to a particular question undergo substantial amounts of editing, this suggests that respondents may be consistently misunderstanding the question, or when the interview is administered by interviewers, that the interviewers may be making systematic errors. If an interviewer is involved, some measure of interviewer characteristics can be useful context. For example, if experienced interviewers are collecting consistently more "no" responses when a "yes" response would lead to an additional set of questions, this could suggest that veteran interviewers are subtly biasing answers to lighten their burden (and that of the respondents).

Survey ***paradata*** include information about the processes that generated the final individual data records, which also can be taken to include metadata as a subset. The wider categories of paradata can include aspects of individual interviewers' speech, such as whether they read the question exactly as intended, whether they probed for more information; indications of respondent effort or uncertainty, such the response latency or changes to initial answers; indications of the use of auxiliary information by respondents, such as administrative records; case history information on all attempts to interview each respondent; an indication of the mode of data collection; information on imputation; information about interviewer training and support; cognitive evaluations of survey questions; computer routines for data processing and imputation; and other systematic processes affecting the final data.

The collection and recording of metadata varies in difficulty and cost depending on the mode of data collection. In particular, computerization greatly lessens the monetary cost of collecting some types of paradata, particularly records associated with the management of cases and the traces of screen navigation recorded as mouse clicks and other key entries during an interview; up-front programming is typically the only such nonnegligible monetary cost. Such interface actions comprise the vast majority of a respondent's outward behaviors during a web-based questionnaire or an interviewer's behaviors in a computer-assisted interview. Similarly, interviewer and respondent speech can be easily captured digitally and then linked to the associated answer, when respondents can be persuaded to give permission for such limited recording.

The capture of metadata may also be facilitated by computerization as well, because it is a simple matter to merge conditional question wording or interviewer information (already entered for a data collection session) with the answers. In addition, the main data and their original state are known with certainty without the intervention of coding and subsequent entry processes that may introduce additional error.

Although a paper-based system of data collection may be made to yield some of the same information as more fully electronic systems, the necessary data linkages are often more difficult, and such linkages allow the possibilities of new sources of error. In such a system, most acts of creating metadata or paradata are inherently costly and thus obvious candidates for omission in a world of continuing cost control.

Without at least basic metadata and paradata, it becomes difficult to find a well-founded empirical basis for evaluation and improvement of a survey under actual field conditions. Without such information, there is only the informal (but clearly very important) knowledge embedded in the actors in the data collection process. Analysis of systemic information may often be difficult or inconclusive, but it is the best hope for informing analysts about the quality of the data and the data collection process and for plotting future improvements.

Collection and preservation of metadata and paradata in the ARMS program appears minimal and unsystematic.[4] A key example is the fact that ARMS is not set up in a way that allows for preserving a record of the original data. Editing and imputation of various sorts occur in ARMS at different levels of data processing, and the information about the outcomes is not systematically stored apart from the data. To gain a sense of

---

[4]Several components of the publicly available ARMS documentation may be considered to provide metadata (http://www.ers.usda.gov/Data/ARMS/GlobalDocumentation.htm). The ARMS data page (http://www.ers.usda.gov/Data/ARMS/) provides a complete variable listing (http://www.ers.usda.gov/data/arms/Variables.htm). Also, the ARMS tailored report query tool has help buttons with definitions (http://www.ers.usda.gov/data/arms/app/Farm.aspx).

the reliability of an observation, it is important that data users be given a clear sense of how much manipulation has been made to the original data; at a minimum, a strange-looking data value might be more credible if it had an associated data flag indicating that it had been reviewed with the respondent or had been reconciled with other variables in the record. Although users may very often want to benefit from the expertise of people who process a large technical database like ARMS, they should have the opportunity to make their own decisions about how to treat data.

Data users have a need for tracking and understanding the impact of imputations for missing data. The relatively simple conditional mean imputation practices used for much of ARMS will generate data that are not appropriate for sophisticated multivariate analysis; for such work, users would need to perform their own imputations or use techniques for analyzing partially observed data. Within the metadata framework, ERS can provide signals to users regarding what values were reported and what values were imputed.

Another critical area in which such knowledge is not recorded in a usable form in ARMS is the history of management of individual survey cases. Systematic collection and organization of such data on attempted contacts with respondents, together with relevant data on interviewer, respondent, and neighborhood characteristics, are particularly important for use in understanding and potentially improving the methods of case administration as well as in understanding nonresponse and detecting nonresponse bias. In light of the relatively high nonresponse rate in ARMS, making such data available should have a high priority.

Highlighting these two issues in no way should be taken to diminish the importance of collecting and organizing other metadata and paradata. In particular, efforts should begin to collect systematic information on interviewers, to document the processes underlying questionnaire design, to document changes in interviewing practices, to note the types of records respondents use, to record any special efforts or incentives used in gain the cooperation of the respondent, and other such factors.

As ARMS moves toward computerization—a step we advocate in this report and that seems inevitable in the long run—it makes sense to build the capabilities for capturing and organizing metadata and paradata as an integral component of ARMS data collection, processing, and products. The need for metadata and paradata makes the transition to digital data collection much more urgent.

Recommendation 5.5: NASS and ERS should develop a program to define metadata and paradata for ARMS so that both can be used to identify measurement errors, facilitate analysis of data, and provide a basis for improvements to ARMS as part of the broader research and development program the panel recommends.

6

Nonresponse, Imputation, and Estimation

In this chapter, the panel explores several important aspects of survey management and methodology that have critical roles to play in either contributing to or minimizing the error in the estimates. Nonresponse, imputation, and estimation are, in the view of the panel, interrelated. The high level of nonresponse in the Agricultural Resource Management Survey (ARMS) triggers the need for imputation of missing values. Imputation is an important initial step in generating estimates. Each of these processes can be a source of nonsampling error.

UNIT NONRESPONSE

When a sample unit fails to respond to the survey solicitation (unit nonresponse) or fails to complete an item on the questionnaire (item nonresponse), it may diminish the representativeness of the sample and thus lead to bias. Typical for large-scale sample surveys, ARMS experiences substantial unit nonresponse. Nonresponse is readily quantifiable, and, possibly as a result, the National Agricultural Statistics Service (NASS) has concluded that unit nonresponse is the survey's biggest source of nonsampling error.[1]

In this section we address methods for reporting ARMS nonresponse, consider the nature of both unit and item nonresponse in ARMS, and discuss methods for reducing nonresponse and making nonresponse adjustments in ARMS.

[1]Statement by Bob Bass, NASS, September 28, 2006.

TABLE 6-1 ARMS Response Rates, 2002-2006

Survey	Year	Sample Size	Response Rate[a]
ARMS screening[b]	2006 *	80,000	77.0
	2005 *	60,000	77.0
	2004	73,376	76.6
	2003	16,638	74.5
	2002	49,156	76.9
ARMS Phase II production practices	2006 *	5,500	80.9
	2005 *	5,500	80.9
	2004	4,755	80.6
	2003	8,148	79.5
	2002	3,421	79.2
ARMS Phase III cost and returns	2006 *	30,000	72.0
	2005	34,203	70.5
	2004	21,075	67.7
	2003	33,861	62.8
	2002	18,219	74.0

[a]Includes operations that responded but were out of scope.
[b]ARMS screening identifies operations that have target commodities for Phase II and for the vegetable chemical use survey.
*Estimated.
SOURCE: National Agricultural Statistics Service, 2005 (updated by the agency).

Response Rates

As might be expected, the response rates reported by NASS vary among ARMS phases and years, since the size of the samples and the target audience varies from year to year. Table 6-1 shows the most recent reported and estimated response rates. The response rates are highest for the Phase II survey and highest for the Phase III survey in agricultural census years. The noncensus year Phase III response rates are the most troublesome, since they fail to meet the Office of Management and Budget (OMB) threshold of 80 percent, below which the agency must plan for a nonresponse bias analysis (U.S. Office of Management and Budget, 2006a, p.8).

Moreover, these published response rates tell only part of the story. The rates for each phase are calculated independently and are shown in Table 6-1. NASS computes the response rate for each phase, as defined by OMB, that is, the percentage of sample units that was accessible and did not refuse the survey. The following formula is used to calculate the response rate for each phase of ARMS:

$$\text{Response rate for Phase } k = \frac{n_k - Refusal_k - Inacessible_k}{n_k}$$

Where: k denotes the specific phase of the ARMS (e.g. I, II, or III)
n_k is the Phase k sample size
$Refusal_k$ is the count of samples units that refused to respond to Phase k of the survey
$Inacessible_k$ is the count of sample units that were inaccessible (unable to be contacted) during Phase k of the survey

These independently computed rates do not take into account nonresponse from the previous phase(s). For certain uses, this is appropriate. For example, to assess the success of enumerators in a given phase, one needs to know the number of completed questionnaires divided by the number of units that were to be contacted. However, for assessing error in the survey estimates, the denominator should also include eligible units from the original sample that would have been contacted if not for their nonresponse in an earlier phase.

Put differently, in a survey with more than one phase, nonresponse typically cumulates from phase to phase. For example, the 2005 Phase II response rate is reported as 75.3 percent. This calculation appropriately reflects the success of the Phase II survey operation in securing responses, but it does not reflect the fact that the sample comes from the Phase I survey, which had its own nonresponse. Thus, the Phase II response rates reported in the table overstate the proportion of the eligible sample that participated in Phase II (and likewise for the Phase III rates).

NASS calculates but does not publish a cumulative response rate for ARMS. In essence, response rates for sample components that participate in more that one phase of ARMS are adjusted by the response rate for that component in prior phases to arrive at a cumulative response rate for each component. The sample size of the component proportionate to the total sample size for all components in the given phase is used as a weight to composite the cumulative response rates for a phase. To complicate matters, the cumulative response rate for ARMS takes several different computational schemes, depending on the relationship of the particular component to prior phases. The steps in the computational scheme are shown in Box 6-1.

The cumulative response rate for Phase III is shown in Table 6-2. According to the steps shown above, it is computed according to the following formula:

BOX 6-1
Steps To Compute a Cumulative Response Rate for ARMS

Step 1: Adjust each component for prior phases.

Phase III Production Practices and Costs Report—The only traditional Phase III component progressing through Phase II is the PPCR component. Calculate the adjusted Phase III response rate by multiplying the Phase I, Phase II, and Phase III response rates.

Costs and Returns Report (CRR) Components—These components only require adjustment for Phase I. Therefore the adjusted CRR response rates are the product of the corresponding Phase I and Phase III response rates.

CRR NOL (not on list) Component—The ARMS Area component does not participate in prior phases and requires no adjustment.

Step 2: Calculate the proportional weight based on sample size for each component.

Calculate the proportional sample weight by dividing each component's Phase III sample size by the sum of the samples sizes from all Phase III components.

Step 3: Weigh the adjusted response rates.

Calculate the adjusted weighted response rate for each component by multiplying the component's adjusted response rate (step 1) by the proportion of sample weight (step 2).

Step 4: Sum the adjusted weighted response rates.

Sum the adjusted weighted response rates from each component.

$$\text{Cumulative response rate for Phase III} = \frac{\sum_i R_{1i} R_{2i} R_{3i} n_{3i}}{n_{3T}}$$

Where: R_{1i} is the Phase I response rate for component i
R_{2i} is the Phase II response rate for component i
R_{3i} is the Phase III response rate for component i
n_{3i} is the Phase III sample size for component i
n_{3T} is the total sample size across all components for Phase III

TABLE 6-2 Unadjusted and Cumulative Response Rate for 2005 ARMS Phase III

	Unadjusted (%)	Cumulative (%)
Response	70.5	51.9
Refusal	23.7	6.2
Inaccessible	5.8	1.3

The proper portrayal of the true extent of nonresponse for understanding the nature of the problem requires that NASS routinely make available not only a cumulative and unadjusted response rate, but also the information needed to independently compute response rates across phases of the survey. This involves showing the disposition of all cases, particularly how the cases in Phase II and Phase III trace back to Phase I. For example, some cases are not in scope for the survey, either out of business or not producing the commodity of interest for Phase II. In tracing these case dispositions, NASS should develop a set of categories that reflect the ARMS survey, but also that take into consideration the categories specified by the American Association for Public Opinion Research and other sources (2006; see also Hidiroglou et al., 1993). Given their differential selection probabilities, this information should be presented separately for the 15-state oversample and the remainder of the sample.

Recommendation 6.1: NASS should routinely report ARMS case dispositions linked across survey phases to provide the foundation for appropriate response rate calculations for Phases II and III.

We also note that interviews from the 15-state oversample carry smaller weights in the production of national estimates and thus should contribute correspondingly less to the overall response rate. A weighted response rate is therefore the most appropriate one to report. NASS would increase an understanding of the extent of nonresponse with this information to report weighted response rates for each survey phase, appropriately reflecting the nonresponse from the preceding phase(s).

Recommendation 6.2: All published ARMS response rates for Phase II and III should be calculated to reflect the nonresponse from the preceding phase(s).

Nonresponse Error

As important as it is to get the response rate right, the response rate is not a measure of nonresponse error, despite its often being used as if it were.

Nonresponse error is a joint function of the proportion of nonrespondents and their distinctiveness. Thus response rates measure only the *potential* for nonresponse error (Groves, 2006).

Although the potential nonresponse error increases as the nonresponse level grows, recent research has found that actual nonresponse *bias* may sometimes be unaffected by increases in the nonresponse rate. Keeter et al. (2000), Curtin et al. (2000), and Merkle and Edelman (2002) found little, if any, connection between nonresponse rates and nonresponse bias, and Groves (2006) reported only a small association between nonresponse level and nonresponse bias in a meta-analysis of studies that had validation measures. These findings suggest that at least some of the characteristics measured in surveys are either uncorrelated, or only weakly correlated, with the causes of nonresponse (Groves et al., 2004). Other characteristics, of course, will be more strongly associated with nonresponse. It is generally impossible to judge a priori which outcome is more likely to occur.

Unlike unit nonresponse rate, which is a characteristic of a survey, unit nonresponse error is a feature of a survey estimate. The same survey may generate estimates containing widely varying nonresponse errors. For example, the ARMS estimate of farm income could have serious nonresponse bias at the same time that its estimate of pesticide use had no nonresponse error. Moreover, even if the ARMS estimate of the univariate distribution of farm income did have nonresponse bias, this would not mean that the ARMS estimate of the multivariate farm income distribution (say, how income is related to acreage and pesticide use) also had nonresponse bias. Just as there is no necessary connection between the nonresponse errors of different variables in the same survey, there is no necessary connection between the nonresponse errors associated with descriptive uses of a survey (involving a focus on means and totals) and the nonresponse errors associated with analytic uses of the same survey (involving a focus on model coefficients).

As far as we know, attention in ARMS has focused largely on the nonresponse rate, as opposed to nonresponse error. A shift in emphasis is therefore warranted. Research should focus on the nature of the ARMS nonresponse bias. If call-record histories are available, an inexpensive first step is to run simulations of the impact of reductions in response rate from the existing level. That is, estimates from ARMS can be compared with estimates from the same survey deleting the last (and probably most expensive) 5 percent (or 10 percent, or more) of the cases. (If call-record histories are not available, the survey should change its archiving practice for the next round of data collection.)

Given sufficient uniformity in the management of fieldwork, these simulations would provide a rough approximation of the impact of various increases in unit nonresponse from the current level. Equally important, the results could provide clues to the effects of various levels of nonresponse on estimation bias. If "last" cases are more similar to actual nonrespondents in ARMS than to other respondents, this procedure could give guidance to field management strategies that would reduce overall nonresponse bias. Another approach might be to use a nonresponse follow-up survey—a sample of nonrespondents—to obtain at least summary information on key characteristics of nonrespondents. Interviewer observations of nonrespondents and capture of detailed reasons given for not participating might also provide valuable insights. In light of the high nonresponse rates in ARMS, the project should be using a variety of methods to understand the nonrespondent population.

Without this kind of information, researchers make survey design changes aimed only at increasing the overall response rate (or stemming the overall decline in response rate). But in terms of data quality, decreasing bias is generally viewed as relatively more critical than increasing statistical efficiency through obtaining more completed interviews, although efficiency is, of course, important. If it is possible to raise the response rate close to 100 percent and cost is no object, then maximizing the response rate is the best strategy. But if this is not possible and if sufficient information is available, then efforts to target certain subgroups are likely to lead to greater bias reduction than those targeted at the entire sample. It is possible that strategies to reduce the overall nonresponse rate could even increase nonresponse bias, if they are disproportionately effective at increasing the response of groups that are already overrepresented. Strategies for reducing nonresponse bias (as opposed to those for reducing the overall nonresponse rate) can be cost-effective when there is information about the composition of the nonrespondents. Although such information is unlikely ever to be detailed enough to enable survey managers to target respondents to reduce nonresponse bias, collection and analysis of such information should be an important and routine part of the survey operation in developing cost-effective survey management protocols.

Recommendation 6.3: The nature of the ARMS nonresponse bias should be a key focus of the research and development program the panel recommends. This research and development program should focus, initially, on understanding the characteristics of nonrespondents.

Methods for Reducing Unit Nonresponse

Despite the preceding discussion, it is very likely that reduction of nonresponse in ARMS should remain a priority, albeit one more informed by study of survey processes and what can be learned about nonrespondents. Because nonresponse at the high rates observed in ARMS introduce substantial uncertainty about the nonrespondent population, the program should also remain mindful of the importance of addressing the concerns of sample members that lead to nonresponse.

All responses in ARMS are voluntary. Thus, establishing interest among respondents and enhancing the survey's credibility may be important in improving response rates. The U.S. Department of Agriculture (USDA) publishes advertisements in trade journals and preinterview letters are sent to sample members. The outreach program solicits active cooperation of commodity interest groups and other private organizations, and, in a special pilot study, a one-hour program was produced and televised nationally in September 2004 and February 2005 to promote the 2004 ARMS. In all, 11 percent of respondents completing the 2004 ARMS Phase III economic enterprise (face-to-face) version indicated that they had seen the program (National Agricultural Statistics Service, 2005, p. 22). Another one-hour program was produced and televised during similar time frames for the 2005 ARMS. The panel thinks that further work along these lines is merited, and it would be enhanced by more rigorous evaluation efforts.

Timing of the phases of the survey may also affect survey cooperation. Part of Phase II occurs during the busy harvest season. Phase III occurs in the spring, typically before the busy planting season, which may impose less of a burden on operators' schedules. This timing may also be well suited to minimizing recall errors (e.g., requesting information on revenues and expenses in early spring is advantageous because many operators will have recently completed their tax returns). As far as we know, there is no systematic research that establishes how survey timing is related to ARMS data quality. Thus the design and interpretation of the research and development we have recommended on both nonresponse and measurement might consider the issue of survey timing.

The survey has also experimented with the use of incentive programs to increase response (as well as response quality). Incentives research was conducted in 2004 on the ARMS Phase III core questionnaire that is mailed to respondents. The incentive was a $20 ATM debit card. NASS also offered a $20 debit card with the spring phase in 2005 to 16,000 sample units. The overall percentage of households cashing the card was 33-35 percent, or 40 percent of respondents and 3 percent of nonrespondents. Of those who got the card after responding to the survey, the rate was 60 percent. Offering the card led to a 6-7 percent increase in response rates for the mail returns. In addition, in 2006 NASS plans to provide customized data products to some respondents as a postincentive. For sufficiently complete returns, a brief report will be provided comparing the respondent's responses with the published summary responses for the respondent's geographic area.

Recommendation 6.4: The research and development program should continue NASS's work on both public relations and incentives, and it should do so with a focus on nonresponse bias, not simply nonresponse rate.

One far-reaching potential change to improve nonresponse has not been employed: that is, to make ARMS mandatory. Since 1997, ARMS and the Census of Agriculture have been integrated during census years (years ending in 7 and 2) to reduce respondent burden. ARMS content is reduced during census years to facilitate the integration. There were also changes in question wording to provide more consistency between the census and ARMS, which has allowed certain census variables to be refreshed annually with ARMS. The response rate is better in census years. This may be due to the fact that the census is mandatory, and the farms selected for both ARMS and the census in the census years may tend to treat ARMS as mandatory.

The heightened response in years in which the content is reduced and respondents may believe the survey to be mandatory affords the possibility of comparing ARMS results in census years with those in noncensus years, and, depending on the findings, to consider whether action should be taken to make ARMS mandatory, as are the annual Census Bureau economic surveys. Comparison of response rates within groups in the census and noncensus years may also lead to a greater understanding of potential sources of nonresponse bias.

Recommendation 6.5: The research and development program should analyze whether there are differences in ARMS unit and item nonresponse rates between census and noncensus years, with an eye toward deciding whether making ARMS mandatory would improve data quality.

ITEM NONRESPONSE

According to NASS, item missing data varies across the variables measured in ARMS. Table 6-3 shows the distribution of variables in the 2004

TABLE 6-3 Percentage Distribution of Variables by Missing Data Rate, ARMS Phase III, 2004

Refusal/Unknown Rate	
10% or more	4.1
5 to 9.99%	11.5
1 to 4.99%	26.0
0 to 0.99%	58.4
All	100.0

Phase III survey by missing data rate. To support this summary table, we were provided a considerable amount of response data by variables (item codes). However, response rates by variables and characteristics of respondents were not available due to confidentiality concerns.

Many of the items with the highest rates of nonresponse involve dollar amounts for such things as assets, income, and debt. There is evidence from other surveys that this problem can be minimized through questionnaire design approaches, such as using respondent-generated intervals rather than point estimates (Press and Tanur, 2000; Juster and Stafford, 1991). These approaches should be a focus of the research and development program the panel recommends.

As noted in Chapter 5, there is concern that the ARMS questionnaire is quite complex and burdensome. One effect may be that respondents might provide false answers in order to avoid additional questions or that interviewers might steer respondents or fill in responses on their own. If such behavior occurs, it could represent an important latent form of nonresponse. This possibility further underscores the need for the ARMS program to do systematic research on how the questionnaires are perceived, how they are administered, and how they are answered. To some degree, such behavior can be detected by including consistency checks within an interview.

Recommendation 6.6: The research and development program should examine how questionnaire design and interviewing changes could reduce item nonresponse

IMPUTATION

When variables are imputed, originally missing data are replaced with values generated by an assumed model for the nonrespondent data, so that analyses and summarization can more effectively be performed. The most common methods that statistical agencies employ for imputation are based on replacing missing values with reported values from other units in the survey, or mean values for that variable for respondents in a similar group, or values generated by other model-based methods. Imputation is useful, even necessary, to support analysis and summarization, but if it is improperly done, imputation can introduce nonsampling error and attenuate measures of association among variables.

The imputation processes discussed in this section are not to be confused with changing data that are regarded as erroneous. Errors, which are typically discovered using logical comparison programs, usually reflect conflicts between the responses on a record: for example, a record reports no spouse but reports earnings by a spouse; or a record reports that 1,000 hogs in total were removed from the operation in Section C, but also reports that 10,000 hogs were removed under a production contract in Section D. As in these examples, the inconsistencies may reflect simple transcription errors. NASS statisticians are careful about identifying and correcting errors—that is, conflict doesn't necessarily imply error, and survey forms and notes are closely checked to provide further evidence, before any changes are made (in the examples above, other responses, such as age or ethnicity queries for the spouse or payments received for the hogs, give further clues as to the correct entry). Errors may be corrected manually, but the process should not be confused with imputation procedures.

Imputation is a more formal process in ARMS and is a shared responsibility between NASS and the Economic Research Service (ERS). ARMS survey items are divided into two categories—those for which a response must be provided (either by the respondent or by a NASS statistician) and those that can be initially coded as a nonresponse. For a questionnaire to be "complete," respondents must provide physical data on acreage, production, and farm production expenses. These items are not subject to imputation. For the items that must have a response, a NASS statistician typically uses manual methods to input data when the respondent can't provide an answer.

About 75 percent of ARMS Phase III survey items fall in the category of items that can be individually refused. Refusals are designated with a minus 1 (–1) entry in the record. Using the 2005 version 1 Cost and Returns Report (CRR) questionnaire as an example, 523 items could be individually refused.

In order to meet mission requirements for developing annual farm business and farm household financial estimates, ERS designates a set of items for which imputed values must replace item refusal codes on completed questionnaires. Some of the imputations are performed by NASS and some are performed by ERS. In general, NASS creates imputations that can be based on mean reported values for farms that fall into the same location, farm size, and farm type categories as the refusal. ERS performs more complex imputations for farm and household financial refusals. The division of labor also reflects program needs: in general, NASS doesn't produce statistics for farm households, whereas ERS does, so ERS develops the needed farm household imputations.

At the request of ERS, NASS employs model-based imputation data for a small subset of the items that can be initially coded as nonresponses. The remaining items remain coded (using –1 as a value) as nonresponses. Again using the 2005 CRR example, ERS asked NASS to replace refusal codes with imputed values for 100 of the 523 refusable items (the "cost of production" ARMS versions 2-4 add many more refusable items, and the total number of refusable items across all versions exceeds 1,400). The NASS imputations are concentrated in three sections: 27 in Section E (other farm income), 15 in Section G (farm assets), and 44 in Section I (farm management and use of time), on which refused items on labor hours allocations (items 25 and 26) are imputed.

In the research databases, items that cannot be refused and items that are imputed by NASS are coded with a "P" prefix along with the cell number[2] printed on the response box in the questionnaire. Thus, P508 in 2005 refers to cell number 0508 on the survey (cash or open market sales of hogs or pigs), and the P designation indicates that there are no nonresponse codes for that item. Other items, which could have nonresponse codes in the data delivered to ERS, are coded with an "R" prefix. Thus, R1241 refers to cell number 1241 (the year in which the principal operator began any farm operation), and the R indicates that the survey statistician can enter a –1 if the respondent doesn't know or refuses to provide an answer.

ERS imputes values to replace refusal codes for about 40 of the remaining refusable items. These items are necessary to produce farm business and farm household financial estimates. One item is in farm assets (Section G) and 3 are in farm debt (Section H). The remainder are in Sections I (farm management and use of time) and J (farm household), with the bulk in Section J. They cover questions on off-farm income earned by the operator household, operator household and family living expenses, nonfarm assets and debt held by the operator household, and operator and spouse characteristics (age, gender, education, and occupation). Those items are redefined as named variables, and the research database includes the named variables as well as the original R-prefixed items. ERS also creates other named variables, and some may be combinations of survey responses (P- and/or R-cells). For example, "total off-farm income" is a named variable that is a combination of cells, and so is "farm business income." Some of those named variables may have an imputed R-cell as a component.

To summarize, about 75 percent of the survey items can be coded with refusal codes. Nonresponse is imputed on about 13 percent of those items, 10 percent by NASS and 3 percent by ERS. Items that are subject to imputation by ERS are renamed as "named variables," with the original data retained in the database as "R-cells." ERS then creates more named variables, usually from combinations of items, and many of these may have imputed data in them.

For the first rounds of imputation, NASS employs both manual and model-based imputation. Manual imputation is used when a questionnaire

---

[2]The cell number corresponds to the item code printed inside or adjacent to the response box. The cell number or item code is used for key-entry purposes. After the data are keyed, the data are placed in a data file. In the excerpt above, fields in the data file are given variable names (labels) based on the cell number corresponding to the response box that was the source of the data value.

is partially completed and may be completed using the knowledge of the state's agricultural experts in the field offices, based on whatever information they have about the farm operation and the knowledge they have of local and market conditions from reported data from other recent surveys or data from other sources. Examples of data from other sources are (a) the average of amount charged or paid by the corresponding contractor for similar arrangements as reported by the contractor, (b) corresponding local averages from other government or industry sources, or (c) data from state–law based mandatory reporting or Farm Service Agency applications.

In their model-based imputations, NASS uses an algorithm that assigns values based on the mean value of valid sample values with reported data from respondents identified as "donors," conditional on a small set of categorical variables, including region, state, sales class, farm size, and type of farm. If there are fewer than 10 donors in a group, then a fallback group is used. (In 2005, a fallback group was not necessary for 77.8 percent of missing data.) When appropriate, a relationship between two items is used to improve imputation—for example, categories of reported owned acreage are used to impute the value of owned land.

Some variables are more likely to be imputed by NASS than others. The majority of imputed items can be attributed to three groups: farm labor, other farm income, and farm assets. The most common imputed data cell is depreciation expenses, which was imputed 2,757 times. In total, NASS imputed 40,253 values (7.6 percent of the total number of nonzero values) in 2005.

ERS responsibility for imputing data items focuses on items that have been identified as important to meet its mission requirements for the annual development of farm operation/farm operator household financial and structural characteristics. These imputations are also selected on the basis of the value of the financial and structural variables that are required for the creation of the ERS ARMS Phase III research file. The methodology for imputing these items is discussed below.

ERS can be said to impute data by using "like" information in a broad sense. But determining what constitutes such information is particularly complex for the ARMS Phase III questionnaire, in which questions and the definition of the data items may vary from version to version of the survey.

Sometimes ERS aggregates input variables to the imputations differently from the underlying data, and separate models must be developed for different versions of the questionnaire. For example, off-farm income from the collective category of "disability, retirement, Social Security, unemployment, Veteran's benefits, other public retirement and public assistance programs" is collected separately for the farm operator, that person's spouse, and other family members in version 1 of the questionnaires. In all other versions of the questionnaire, income in this category is reported for the operator's household and not separately by type of household member. Because of these differences, imputation of the total amount of this income component takes place in two steps. ERS imputes the missing values of retirement income for each of the operator, spouse, and other household members in the version 1 using average values calculated by age and education classifications of the primary operator. ERS also imputes missing values of the total amount of household income in this category in the other versions. Imputation for versions 2-5 is based on data from all versions, since the conditional means needed for imputation require the use of weights that are available only for observations in version 1 and the set of all observations. At other times, ERS imports data from other parts of the questionnaire to aid in imputation. For example, to impute the amount of debt for farm operation loans in three categories (short term, real estate with a term over one year, and uses other than real estate with a term of over one year), ERS uses information reported in the farm debt and farm expense sections of the survey. On version 1, debt is collected by lender while on the remaining questionnaire versions it is collected in the three required categories. Debt on version 1 is first combined into the three categories. A category (for example short term debt) is classified as a refusal if debt is refused for one or more lenders in that category. When one or more categories of debt is/are refused total debt is imputed based on total reported interest expenses (real estate plus nonreal estate). Debt by category is then computed (by farm size and type) for reporting farms. This information is then used to divide total imputed debt into the three required categories.

These complex, individualized, and somewhat judgmental imputation schemes apply to several other key items, which are handled in a special manner due to structural differences in the versions of the questionnaire:

- Off-farm net income from any other farms, income from any other business, and income from renting farmland to others.
- Off-farm income from interest, dividends, and other off-farm income.
- Total proceeds from the sale of farm and nonfarm assets and recognized gain or loss on the sale of capital.
- The number of persons who lived in the primary household as of December 31.
- The components of nonfarm assets, nonfarm debt, and consumption expenditures of the operator household.

There are some other potential problems with the imputation practices used in ARMS. Imputation relies on the use of conditional means as estimates of missing values. For survey estimates of simple univariate-level statistics or such statistics cross-classified by a couple of variables, such an imputation process should be generally adequate. However, estimates of variability in the data will usually be artificially reduced. Except when actual values have an exact relationship with variables used to condition the mean estimate, the true values of the missing data should have a distribution of values, rather than the simple set of imputation values. Moreover, when more complex multivariate relationships are estimated, this approach to imputation may lead to estimates that are statistically inefficient at best and biased at worst. The reason for these problems is that conditional mean imputation generally cannot condition on a sufficiently large set of variables to maintain relationships between the variables imputed and all variables that might be included as related variables in a multivariate analysis. A more broadly conditioned and explicitly model-based approach to imputation would preserve important relationships. Such an approach would also make it simpler to create multiple imputations, which would provide a straightforward means for estimating the degree of uncertainty in the ARMS estimates that are due to missing data.

In judging the adequacy of the current ARMS imputation procedures, the panel refers to the following criteria (Charlton, 2007):

- *Predictive accuracy.* The imputation procedure should maximize preservation of true values. That is, it should result in imputed values that are close as possible to the true values.
- *Ranking accuracy.* The imputation procedure should maximize preservation of order in the imputed values. That is, it should result in ordering relationships between imputed values that are the same (or very similar) to those that hold in the true values.
- *Distributional accuracy.* The marginal and higher order distributions of the imputed data values should be essentially the same as the corresponding distributions of the true values.
- *Estimation accuracy.* The imputation procedure should lead to unbiased and efficient inferences for parameters of the distribution of the true values (given that these true values are unavailable).
- *Imputation plausibility.* The imputation procedure should lead to imputed values that are plausible. In particular, they should be acceptable values as far as the editing procedure is concerned.

In the panel's view, although the current methodology for imputing with partial nonresponse may satisfy the first criterion for means, totals, and perhaps for ratios, it can lead to poor results when the analyst is interested in distributional properties of the population.

Another concern was brought to the panel's attention by members of the research community who use the ARMS data. Good statistical practice is to identify when data have been imputed. The Federal Committee on Statistical Methodology has argued that, at a minimum, data users should be able to identify the values that have been imputed and the procedure used in the imputation. This transparency permits users the option to employ in their analyses their own methods to compensate for the missing values (Federal Committee on Statistical Methodology, 2001, pp. 7-8). This is particularly important when the individual record data are made available for research and other purposes. Failure to flag imputed data may affect the reliability of variables derived from the databases, and it will cause users often to overstate the statistical significance of their results

ARMS fails to identify the imputed data, but it is possible to ferret out the imputed from nonimputed values with a bit of work. Since the original R-cells are retained on the research database, those with an indicator of –1 are refusals. The associated named variable includes the imputed value, so a researcher can identify when an imputation has been made and can also identify the imputed value, by comparing R-cells with associated named variables. In order to do so effectively, external researchers need to be provided with much more guidance on ERS imputation procedures and a concordance that links named variables to those R-cells that are subject to imputation by ERS. Another option would be to simply flag the imputed values.

Recommendation 6.7: NASS and ERS should consider approaches for imputation of missing data that would be appropriate when analyzing the data using multivariate models. Methods for accounting for the variability due to using imputed values should be investigated. Such methods would depend on the imputation approach adopted.

Recommendation 6.8: All missing data that are imputed at any stage in the survey should be flagged as such on files to be used for analysis.

ESTIMATION

The estimation method is one of the determinants of the size of the error in the estimates. Generally speaking, statistical agencies seek to employ estimation methods that result in both the smallest bias and the smallest sampling variance (Statistics Canada, 2003). However, due to its unique design, some of the standard estimation methods are not available for ARMS. For example, use of estimation methods that exploit the correlation over time in periodic surveys, with a large sample overlap between occasions so that data from previous rounds can be used as auxiliary variables (*composite estimation)*, is not possible in ARMS because of the lack of sample overlap.

Because of informational constraints and institutional traditions, the estimation of ARMS statistics at NASS appears to be quite unconventional. ARMS estimation is a carefully choreographed multistage process that involves multiple inputs, including extra-survey inputs.[3] This process derives from a long-standing tradition at USDA to develop estimates based on collected information filtered through a board process that applies the expertise of NASS and ERS staff who are selected to serve as members and their knowledge in the generation of the official estimates. The statistical estimates computed directly from ARMS, which in other government surveys might be considered the final estimates, are treated as "indications" in a larger process of estimation. The process is intended to maximize the use of what is believed to be the best information and to ensure consistency with other estimates that are published by USDA.

The estimation process at NASS begins rather conventionally with the stages of imputation and editing followed by estimation (called summary activities). It then begins to take a path that is distinct for USDA surveys, proceeding along the following steps:

1. Traditional nonresponse adjusted summary.
2. Calibration and preliminary calibrated summary
3. Outlier identification and replacement (Outlier Board)
4. Rerun calibration and summarization incorporating outlier replacement (Final Calibrated Summary)
5. Establishment of farm production expenditures estimates at the national level (U.S. Farm Production Expenditures Board)
6. Establishment of farm production expenditures estimates for 5 farm production regions (Regional Board)
7. Establishment of state-level farm production expenditures (State Board)
8. Establishment of final estimates for economic class, farm type, and fuel subcomponent expenditures

Summary. During summary, variances are computed using the delete-a-group jackknife method. The output from the summary is called an "indication" rather than an estimate.

Two summaries are created: a traditional nonresponse-adjusted summary and a calibrated summary. For the former, traditional nonresponse adjustments (reweighting) are made and direct expansions are computed for all items, including the production items. The expanded number of farms from the ARMS is compared with the NASS official estimates for farm numbers to evaluate the level of representation of the responding sample. Weights are scaled using the official farm number estimates as targets.

[3]NASS "Estimation Process" presentation given at the ARMS III Workshop, July 2006; and NASS "Estimation Manual," 2006.

However, for the calibrated summary, no traditional response adjustment is applied prior to calibration, since calibration is presumed to adjust for nonresponse as well as coverage.

Traditional nonresponse summary indications and preliminary calibrated summary indications are reviewed by the NASS field offices. When outlier adjustments are recommended, documentation is provided to the National Expenditures Board to use in determining whether to make a change.

Calibration and Preliminary Calibration Summary. The ARMS estimation process actually begins with the calibrated summary. "Calibration" is a general term for a sampling-weight adjustment that forces the estimates of certain item totals based on the sample at one phase of sampling to equal the same totals based on a previous phase or frame (control) data. As employed by statistical agencies around the world, calibration is a common and accepted procedure for adjusting for unit nonresponse and for undercoverage. NASS includes a sample of nonoverlap (NOL) farms in the original sample as discussed on Pages 4-5 and 4-6. This provides a direct measure of list incompleteness for all survey indications.

Since the 2004 ARMS a calibrated summary had been produced. A multivariate calibration has been performed which uses official estimates from other survey sources (14 listed items) to assess the level of representation of the responding sample. At that time, the decision was made to confound the nonresponse adjustment with the coverage assessment. The term "coverage" may or may not include list incompleteness and can be overcoverage as well as undercoverage. In ARMS, the NOL farms address list undercoverage, but they are eligible for calibration adjustments.

At this point, NASS adds another step to meet another goal. In the process of calibration, NASS uses *truncated*[4] linear calibration to meet predetermined official estimates for acreage and production for specific items raised on the farms, called calibration targets (see list below). The ARMS weights are adjusted so that the ARMS survey replicates the control totals for these 14 items. The expenditure estimates are then regenerated with these "calibrated" weights. This calibration process ensures that the ARMS data replicates the official NASS estimates for the 14 selected crop and livestock items set by the board process.

The following official NASS estimates are used as calibration targets:

1. Corn for grain, harvested acres
2. Wheat for grain, harvested acres
3. Soybeans, harvested acres

[4]Calibrated weights are constrained to be greater than or equal to one, in the belief that each respondent should at least represent itself. Other constraints may be applied as appropriate.

4. Cotton, all, harvested acres
5. Potatoes, harvested acres
6. Vegetables acres (sum of fresh and processing)
7. Fruit and Nut acres (sum of 28 fruits and nuts)
8. Broilers raised
9. Turkeys raised
10. Eggs produced
11. Cattle inventory
12. Pig inventory
13. Milk production
14. Number of farms (by 7 economic sizes)

Outlier Identification and Replacement. The total calibrated expenditures are used to determine objective decision rules for identifying outliers. These are referred to the respective Field Offices for closer examination. Field Offices check for undetected non-sampling errors, review the history of the respondent, and develop a more detailed profile of the operation. This information is used by the Outlier Review Board to decide on a consensus course of action. These actions are posted to the dataset and the same calibration procedure is rerun with these additional constraints using the 14 external targets listed above.

Review of Final Summary. After calibration, the calibrated summary is rerun. These regenerated indications are delivered to the National Expenditures Board and the field offices. All field offices again review the indications and write comments, which may be presented to the National Expenditures Board. During this time, NASS and ERS staff prepare for the National Expenditures Board.

Preparation of Official Estimates. The National Expenditures Board sets farm production expenditure estimates. The National Expenditures Board consists of about 20 NASS and ERS representatives, and, since 2006, state board members. In a consensus process, it sets 16 farm production expenditures estimates[5] at the U.S. level, and sets the stage for subsequent establishment of expenditure levels for farm regions, states, and economic class, farm type, and fuel subcomponents. Later, the board sets estimates for the 5 farm regions, and, supplemented by representatives of the states, sets estimates for the 15 largest agricultural states.

The estimation process has four major inputs: estimates from the previ-

[5]Farm production expenditure estimates are prepared for feed, farm services, rent, agricultural chemicals, fertilizer, lime and soil conditioners, interest, taxes (real estate and property), labor, fuels, farm supplies and repairs, farm improvements and construction, tractors and self-propelled farm machinery, other farm machinery, seeds and plants, and trucks and autos.

ous year; ARMS indications from the current year; other data; and knowledge of NASS, ERS, and state economists and statisticians. The data other than ARMS that the board uses in its work include the Census of Agriculture, prices paid indexes from the NASS Agricultural Prices Report, Crop Acreage and Production, Livestock Production, ERS farm income, ERS cost of production for various commodities, and data from the National Pork Producers Council, the Federal Reserve, the producer price index (PPI), and the Association for Equipment Manufacturers (AEM). ARMS III staff prepares recommendations prior to the board.

The boards rely on two major inputs or "summaries." One is a direct expansion of the summarized data. The other is the final calibrated summary.

Regional Estimates. After the national estimate for each expenditure item is set, board members set regional-level estimates for the five farm production regions. These regional estimates are constrained to sum up to the National Expenditures Estimates.

State Estimates. The board members then set state-level estimates that sum up to the regional expenditures estimates. Thus, in a cascading effect, the board conducts several reviews to ensure that the totals estimated in the ARMS estimation process are consistent with official production estimates, and that regional and state estimates are consistent with the national totals and are additive.

Other Estimates. Other values are estimated directly, including the level of aggregate expenditures by item within each size class of gross farm revenue, and within each major type of farm enterprise and fuel subcomponent expenditures (diesel, gas, liquid propane gas, and other fuels).

In this section, we have outlined the several steps in the development of published estimates from ARMS. Overall, the effects of the various adjustments on statistics estimated using ARMS are not clear. In particular, the interventions in ARMS based on board processes introduce changes that are not replicable in the normal sense expected in scientific research. These interventions may well lead to better estimates, or they may simply impose consistency across various key estimates at the cost of disturbing other relationships in the data.

Recommendation 6.9: NASS and ERS should provide more clarification and transparency of the estimation process, specifically the effect of calibrations on the assignment of weights and the resulting estimates.

7

Methods for Analysis of Complex Surveys

The panel was asked to consider the appropriateness of the methods the Economic Research Service (ERS) is using to fit statistical models to data from the Agricultural Resource Management Survey (ARMS). The economists at ERS have been advised by the mathematical statisticians at the National Agricultural Statistics Service (NASS) that they should always take a design-based approach for making statistical inferences by using the survey weights. ERS has asked for more explanation on the appropriateness of such an approach. They would like to have a better understanding of the subtleties of this approach relative to a model-based approach, so they can better explain to the users of their analytic outputs, using sound statistical arguments, why they have adopted this approach. At the same time, ERS scientists are also interested in exploring alternative approaches to analyzing ARMS data, especially when fitting complicated econometric models. Finally, ERS is interested in obtaining specific advice on the suitability of statistical and econometric software programs for analyzing ARMS data.

Ideally, ERS would like to have a "Guide for Researchers" on these topics, which would address the various cases that arise in their work. As their analyses become more sophisticated, their questions are becoming more complex. Such a guide would be useful, but writing it is not a trivial task. The panel decided not to write even a short version of such a manual, as this could lead to inappropriate use of the methods if the details of such concepts as ignorability or how to handle analyses based on small sample sizes are not fully understood. Instead, in this chapter we discuss the general principles of design-based and model-based inference, explain the main advantages and disadvantages of both, and provide some guidance on how such a guide could be written. The panel also did not address the software issue in detail, because this depends critically on the specific analyses being undertaken.

We begin with a short description of data analysis issues. We then discuss the conceptual and theoretical underpinnings for the different frameworks under which statistical modeling and analysis can be performed, as well as their implications for analytical inference. In the final section, we offer some guidance on how a Guide for Researchers for ARMS could be written.

DATA ANALYSIS ISSUES

The complex design of ARMS includes stratification, clustering, dual frames, and unequal probability sampling. Each year, NASS provides survey weights that account for these design features as well as for additional information available at the population level and various nonresponse adjustments (see Chapter 6). NASS has also developed and makes available sets of replication weights to facilitate computation of variance estimates, with the current method based on delete-a-group jackknife replication. The survey weights and the replication weights are provided with the ARMS datasets.

Recommendation 7.1: NASS should continue to provide survey weights with the ARMS data set, combined with replication weights for variance estimation.

An important use of data from surveys such as ARMS is for descriptive inference, in which population-level and domain-level quantities of interest are estimated from the survey data. An example of a population quantity of interest for ARMS is the average amount spent on fertilizer by all farms, while a domain quantity of interest is the average amount spent on fertilizer by all farms with annual sales over $50,000. Estimates for population and domain (i.e., a subset of the population) quantities are computed as weighted sums over the sample using the survey weights. The variance is estimated by computing jackknife replicates as the weighted sums for each set of replication weights and averaging the sum of the squared deviations from the mean over the full sample estimates (see Box 7-1). When targeting unknown simple or narrowly conditioned quantities of interest in a finite population and in medium-to-large domains within the populations, this randomization-based type of estimation, and the associated inference in terms of standard deviation and confidence intervals, is generally accepted as being the most appropriate method for obtaining statistically valid estimates (Rao, 2005). NASS and ERS produce large numbers of estimates of this type each year.

Recommendation 7.2: NASS and ERS should continue to recommend the design-weighted approach as appropriate for many of the analyses for users of ARMS data and as the best approach for univariate or descriptive statistics.

In addition to descriptive inference, ERS staff and researchers in other organizations also use ARMS data in analytical inference, in which econometric models are fitted and inference is made about model parameters. In this situation, the goal of the inference is often not finite population quantities, but rather parameters or related quantities for the underlying postulated model, with the population representing a specific realization from that model. Because ARMS data are obtained under a complex design, the general consensus among statisticians is that analytical inference needs to account for the design (Little, 2004). Not accounting for the effects of the sampling design can cause estimators for parameters of the postulated model to be biased, their associated variance estimators to be biased, or both. As a simple example of the possible bias

BOX 7-1
Formula for the Delete-a-Group Jackknife

Consider the linear prediction model, $y_k = x_k\beta + \varepsilon_k$, where x_k represents a 1 x p dimensional vector of predictors for case k, β represents a p x 1 dimensional vector of unknown regression coefficients, and ε_k represents an unknown error term. With the ultimate goal being to estimate $\sum_U y_k$, we collect data from a random sample of the population and use the above model to 'predict' the y_k values for the unsampled members of the universe. A general class of estimators for this problem is that of weighted totals: $\sum_{k \in S} w_k y_k$ (which can, e.g., represent the sum of the observed sample values and the fitted values for the unsampled items) where the S denotes a sum only over the sampled cases. Given this general estimator, what is an estimate of its variance?

Assume in the following that the sample was randomly divided into 15 mutually exclusive groups denoted by S_g, g – 1, . . . , 15, and we let the complete sample with S_g removed be denoted as S_{-g}. The only random quantity in y_k is ε_k, and so if one could observe the ε_k directly, a delete-a-group jackknife estimator of the variance of $\sum_S w_k y_k$ would be as follows:

$$\left(\frac{14}{15}\right)\sum_{g=1}^{15}\left[\sum_{k \in S} w_k \varepsilon_k - \sum_{k \in S_{-g}} \left(\frac{15}{14}\right) w_k \varepsilon_k\right]^2$$

BOX 7-1 Continued

This can be justified as follows. Ignoring the outside term 14/15, inside the square we have the estimate of the total $\sum_{k \in S} w_k \varepsilon_k$ minus an estimate of the same total derived from the S_{-g} set, but multiplied by 15/14, since S_{-g} only represents 14/15 of the sample. Each of the sum of squared terms then provides an indication of how much variability is in the left-out 1/15 of the sample. The entire expression is NOT divided by 15 (or 14) as in a standard variation estimation because only 1/15 of the sample changes within each square term, and so there is only 1/15 of the variability that one is estimating. Finally, the 14/15 term on the outside of the expression eliminates some additional bias.

The only remaining complication is that we do not observe the ε_k—we can only observe residuals from the regression fit. This is a substantial complication, since the covariates x_i can be unbalanced from one group to another. Without justifying this here, one method for dealing with this complication is to replace the ε_k with y_k in the above expression, and replace the w_k with the following:

$$w_{k(g)} = \left(\frac{15}{14}\right) w_k + \left(\sum_U x_i - \sum_{S_{-g}} \left(\frac{15}{14}\right) w_i x_i\right) \left(\sum_{S_{-g}} c_i w_i x_i' x_i\right)^{-1} (c_i w_k x_k'),$$

where the c_i are terms in the regression computation that account for things like heterogeneous variances and correlations among the ε_i's.

The resulting delete-a-group jackknife can then be written:

$$v_{dag} = \sum_{g=1}^{15} \left(\frac{14}{15}\right) \left(\sum_{k \in S} w_k x_k - \sum_{S_{-g}} w_{k(g)} y_k\right)^2 .$$

(Note that the 15/14 is already defined in the $w_{k(g)}$.)

SOURCE: Kott, "Some Thoughts on the Delete-a-Group Jackknife with ARMS-III Data", based on the presentation to the Panel to Review the Agricultural Resource Management Survey (ARMS), June 6, 2006.

introduced by sampling, consider a population of size N where each element has an unobserved value of interest y_i, and that has a true average equal to μ_y, which is unknown. Suppose that there is a second variable x_i that is positively correlated with the y_i, and the x_i are observed for the whole population. Assume now that the sampling design is such that elements are sampled with higher probability for larger values of x_i, which could be implemented through stratification or by using a probability-proportional-to-size design. Then, if one computes the average of the y_i in a sample obtained following this sampling design, it is very likely that it will be larger than the population average of μ_y, that is, the sample mean will be a biased estimator for the population mean μ_y. Holt et al. (1980) and Pfeffermann (1993) give further examples to illustrate this potential bias due to sampling design, which can occur not only for simple estimators as in this example but also in complex modeling situations.

While it is clear to many researchers that the effect of the sampling design needs to be addressed whenever models are fitted to ARMS data, it is less clear how this should be accomplished in practice. The statistical literature describes a number of alternative approaches for incorporating the sampling design in analytical inference for surveys (Little, 2004). One commonly recommended approach, which is also the one currently implemented by NASS and ERS, is to perform model analyses using the provided sampling weights and estimate the variability of parameter estimators under the sampling distribution. In the case of ARMS, the variance estimation can be done by repeating the analysis with the replication weights provided by NASS and computing the average sum of squared deviations over the full sample.

While this is a theoretically sound approach and has some important practical advantages, its current implementation suffers from a number of specific shortcomings, as outlined below.

EVALUATION OF CURRENT APPROACH

The survey weights and associated replicate weights provided by NASS make it possible for researchers to fit economic models using the large majority of statistical and econometric software programs that can handle weighted analyses, although assembling the replicate-based estimates to compute sampling error often requires more detailed programming. Performing statistical analyses using the combination of survey weights and replication weights provided by NASS is a theoretically valid approach that guards against potential bias caused by the unequal sampling of farms in the sample. The method is straightforward to implement, even if not completely automated in most software packages.

However, as currently implemented, this method also has some significant drawbacks, as identified by ERS researchers and some academic users of the ARMS data sets (Dubman, 2001). The small number of jackknife replicates (currently 15) means that the variance estimator has low degrees of freedom. This implies that the variance estimator can be highly variable in some situations, and, more problematically, that it is potentially inappropriate for complex models with large vectors of parameters. This puts a limit on the types of models that can be fitted and analyzed for the ARMS data. An additional problem is that the replicate weights contain negative values in years prior to 2005, causing many software programs to fail to run or to give erroneous answers. Recent changes in NASS procedures eliminated negative replicates for later years.

Two other issues are of concern regarding the jackknife methodology. First, the specific selection method of the 15 replicates is critical to ensuring that the variance estimator is unbiased. The manner in which this needs to be done is well known and appears to be appropriately applied, but clear documentation of the specific implementation for ARMS is needed. Second, several "adjustments," referred to in Dubman (2001) are made so that the jackknife variance reflects the estimation adjustments (calibration, regression, poststratification) made to the original weights. Again, the principles behind these adjustments are well known, but documentation of the specific implementation used for ARMS is needed.

More generally, there is a concern that the current one-size-fits-all approach for analytical inference for ARMS is somewhat restrictive and does not fully satisfy the diverse needs of researchers. Specifically, it would clearly be desirable to some researchers, both inside and outside ERS, if standard survey software (e.g., SAS "survey procs" such as Proc Surveymeans, Proc Surveyreg, and Proc Surveylogistic) could be used for conducting weighted analyses. Other users might be interested in conducting unweighted analyses that rely more heavily on modeling assumptions while remaining aware of the potential sensitivity of their models to the effects of the design and execution of the sample. For both of these types of users, the information currently provided as auxiliary data with the ARMS data file is likely to be insufficient, because these users need more specific information on the sampling design and on the nonresponse patterns and adjustments applied to the dataset to include in their procedures or models. All of these drawbacks of the current implementation of variance estimation can be overcome.

ESTIMATION STRATEGIES FOR COMPLEX SURVEYS

Concepts of Analytical Inference for Surveys

In survey statistics, the postulated model for the finite population from which the sample is drawn is often referred to as the superpopulation model, with the elements in the finite population treated as realizations from the superpopulation. In analytical studies, particularly economic studies, the superpopulation model is typically assumed to correspond to the model the researcher is interested in estimating. As an example, consider the following superpopulation model:

$$Y = f(\mathbf{X};\boldsymbol{\beta})+\varepsilon \qquad (1)$$

where Y and $\mathbf{X} = (X_1, \ldots, X_p)$ are random variables to be observed through a survey, the function f is assumed to be known, and ε is an unobserved zero-mean error random variable. The vector of parameters $\boldsymbol{\beta} = (\beta_1, \ldots, \beta_p)$ is of analytical interest to the researchers. In many situations, the primary interest is not in the joint model for Y and $\mathbf{X}$, but in the conditional model, that is, the model for Y given values of X, in which case the marginal model for X is not explicitly considered in the analysis and X is treated as fixed. For example, Y and X could be dependent and independent variables in a regression model, respectively, and the vector β then denotes the regression parameters for the relationship between them. The regression model is often linear in practice, but the discussion below applies equally to more complicated regression contexts, such as nonlinear models or generalized models.

For now, we assume that the finite population is composed of N independent and identically distributed realizations from this superpopulation model. We use $U =(1, \ldots N)$ to denote the set of indices for the elements in the population. From this population, a sample is drawn according to a sampling design $p_N(s)$, which assigns a probability to each possible sample $s \subset U$. For a given sample s, the researcher wants to estimate the superpopulation parameters β based on the observations $\{(y_i,x_i):i \in s\}$.

The critical issue in how to estimate β is whether the superpopulation model in (1) is a valid representation for the sampled observations$\{(y_i,x_i): i \in s\}$. In the regression model example, this would imply that the relationship between Y and X is unaffected by the manner in which the sample was selected, or equivalently, that the "unobservable" portion of the model, that is, the ε, does not depend on the sampling design. The situation in which the validity of the model of interest is unaffected by the design is referred to in the statistical literature as an ignorable design (see, e.g., Pfeffermann, 1993), based on the terminology introduced by Rubin (1976). It should be noted that ignorability depends on the relationship between the design and the model, so the sampling design for a survey can be ignorable for one model (e.g., the regression model between Y and X) but nonignorable for another (e.g., the joint distribution model for (Y, X)). When the sampling design is ignorable with respect to the model of interest, then it is in principle appropriate to proceed with statistical estimation and inference methods without using the survey weights.

When the sampling design is not ignorable for a specific model, the full model for the sampled observations is different from the original model for the population, or, in the regression context, the relationship between Y and X is different in the sample and in the population. Hence, the nonignorable design can cause estimators that are based on the original model to be biased. Also, ignoring the sample design information when it is nonignorable could lead to inappropriate estimated variances, confidence intervals, and incorrect p-values in a statistical hypothesis test. For an example of how a nonignorable design can induce bias on the relationship between dependent and independent variables in the linear regression context, see Nathan and Smith (1989). In economics, such selectivity issues are widely discussed, but generally they have been addressed with explicit modeling assumptions about the nature of the selectivity, as in the classic paper by Heckman (1979).

When analyzing a data set, it is often impossible to know a priori whether the design is ignorable or not, so it is sound practice to guard against potential biases in estimation and inference. For specific modeling situations, it is possible to perform statistical tests for violations of the ignorability assumption, and variables related to the design can be added to the model as predictor variables. Although adding covariates clearly changes model (1), it makes it possible to create a new conditional model for which the design is now ignorable, so that model-based inference yields appropriate conclusions. An alternative approach, which is valid regardless of whether the design is ignorable or not, is to explicitly account for the sampling design in estimation and inference of model parameters through design-based (weighted) inference. We briefly describe that approach here.

The sampling weights are obtained from the sampling design, denoted by $p_N(s)$, which is used to draw random samples s from the population. Based on the sampling design, there is an associated inclusion probability π_i for each element in the sample s, defined as

$$\pi_i = \Pr(i \in s),$$

that is, the probability that element i is selected into the sample s under the design. For some multiphase designs, the π_i are conditional probabilities that cannot, strictly speaking, be interpreted as inclusion probabilities, but they can be used as such in weight construction. For most commonly used designs, explicit formulas are available for obtaining π_i (see, e.g., Sarndal et al., 1992; Lohr, 1999).

Fixed finite population quantities of interest, such as the population mean or total, can be estimated based on the sample observations, by using the design information captured by the inclusion probabilities. For any variable z, a sample estimator $\hat{z}_\pi$ defined as

$$\hat{z}_\pi = \sum_{i \in s} \frac{z_i}{\pi_i}$$

is design unbiased for the finite population total $t_z = \sum_{k \in U} z_i$, in the sense that over repeated random samples drawn from the population according to the design $p_N(\cdot)$, the mean of the $\hat{z}_\pi$ is equal to t_z. If the population mean is to be estimated, one can divide the estimator for the population total by N, the number of elements in the population.

For more complicated finite population quantities, it is often not possible to maintain design unbiasedness. However, the principle of using sampling weights $w_i = 1/\pi_i$ in the construction of sample-based estimators continues to apply. Under mild regularity conditions on the population and the sample design, any finite population quantity of interest that can be expressed as a function of finite population totals, say $\theta_N = g(t_{z1}, \ldots, t_{zq})$, can be estimated by a corresponding sample-based estimator $\hat{\theta} = g(\hat{z}_{\pi 1}, \ldots, \hat{z}_{\pi q})$, with $\hat{\theta}$ design consistent for θ_N (see, e.g., Sarndal et al., 1992). As a simple example of a function of population totals, consider a variable y_i (e.g., total amount spent on fertilizers by farm i) and another variable x_i (e.g., acreage planted by farm i). According to the above discussion, the ratio $\theta_N = \sum_U y_i / \sum_U x_i$ (i.e., the average fertilizer application per acre planted in the population) can be estimated consistently by the ratio of sample-weighted estimators

$$\hat{\theta} = \frac{\sum_s w_i y_i}{\sum_s w_i x_i}.$$

The function $g(\cdot)$ defining θ_N can be much more complicated than this simple ratio, and the sample-weighted estimation principle applies equally to both explicit population quantities of interest (e.g., coefficients for a population-level regression function) and implicit ones (e.g., solutions of population-level estimating equations).

The preceding discussion of design-based estimation did not refer to a superpopulation model, and indeed the design consistency of $\hat{\theta}$ for θ_N does not depend on any model assumptions about the finite population (except for some mild regularity conditions, such as existence of limits). In analytical uses of survey data, it is possible to combine the above design-based estimation of finite population quantities with the estimation of (superpopulation) model parameters, by taking advantage of the fact that the finite population elements can be viewed as independent and identically distributed realizations from the superpopulation model. Unlike traditional design-based estimation, there are now *two* random processes determining the statistical properties of estimators: the sampling design and the superpopulation model. In what follows, we use the convention of using a subscript ξ to denote properties with respect to the model, p for properties with respect to the design, and ξp for joint properties.

Considering again superpopulation model (1), suppose that it is possible to define a finite population estimator $\mathbf{B}_N$ for the model parameter β, which could be computed if the complete population U were observed. If the components of the parameter vector $\mathbf{B}_N$ can be expressed as functions of finite population totals similarly to θ_N above, it is again possible to estimate it by a sample-based estimator, say $\hat{\mathbf{B}}$, which is a function of design-weighted estimators, as previously done for $\hat{\theta}$.

Figure 7-1 illustrates the traditional view of survey estimators. For the finite population, a parameter vector $\mathbf{B}_N$ is defined (possibly but not necessarily based on a superpopulation model). A sampling process is used to randomly

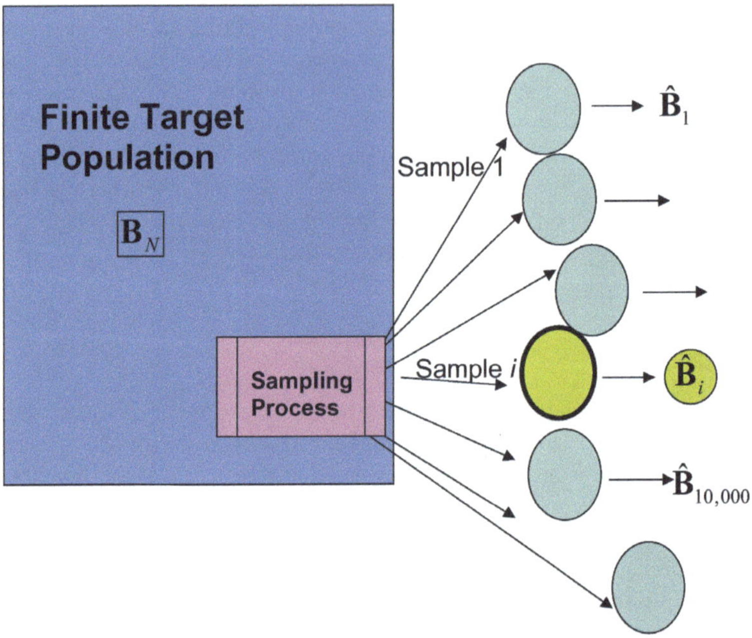

FIGURE 7-1 Diagram representing classical design-based inference.

generate samples, and for each sample an appropriately weighted estimator $\hat{\mathbf{B}}$ can be computed that targets $\mathbf{B}_N$. Under the distribution induced by repeated sampling from the fixed target population, the estimator $\hat{\mathbf{B}}$ will be consistent for $\mathbf{B}_N$, and it is possible to define properties such as $E_p(\hat{\mathbf{B}})$, the design expectation of $\hat{\mathbf{B}}$, and $Var_p(\hat{\mathbf{B}})$, its design variance.

Figure 7-2 takes this idea one step further, by viewing the target population itself as a random realization from the superpopulation model. In this view, the finite population quantity $\mathbf{B}_N$ is viewed as one particular realization of an estimator of the superpopulation model parameter β. From this particular realization, a sample is drawn according to the sampling design, and an estimator $\hat{\mathbf{B}}$ is constructed. By explicitly including the sampling weights in the construction of $\hat{\mathbf{B}}$, we are guaranteed to have a valid (design consistent) estimator of $\mathbf{B}_N$, the finite population "estimator" of β. If the superpopulation model is correctly specified, then $\mathbf{B}_N$ is itself (model) consistent for β. Combining both phases of this estimation process, $\hat{\mathbf{B}}$ is a valid (design and model consistent) estimator of β under joint inference for the sampling design and the superpopulation model.

As noted above, when inference for a model is made jointly under a

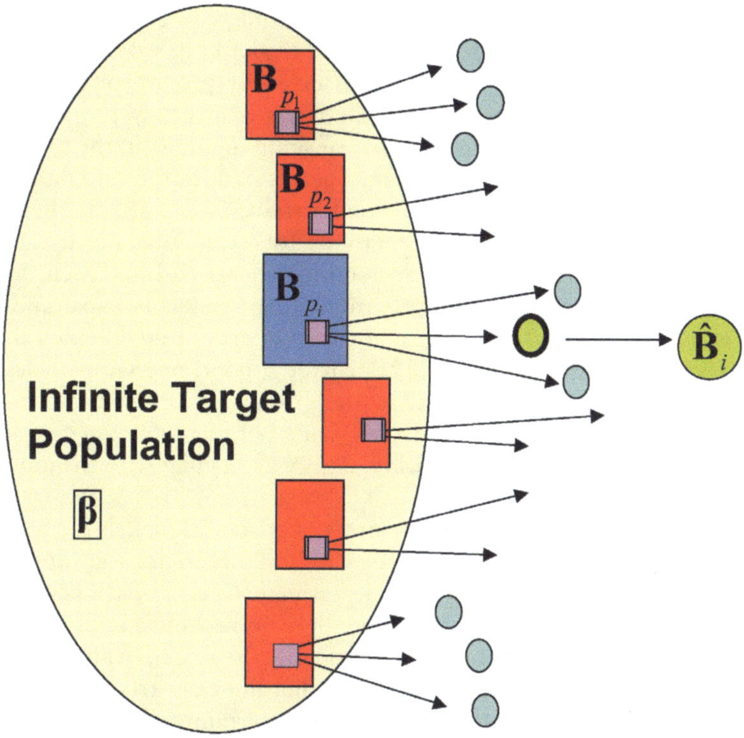

FIGURE 7-2 Diagram representing model design-based inference for superpopulation parameters.

superpopulation model and a sampling design, both of these components contribute uncertainty. Hence, formal inference (confidence intervals, variance, etc.) would need to explicitly incorporate both sources of uncertainty. For instance, using standard conditioning arguments, the joint model-design variance of $\hat{\mathbf{B}}$ for the parameter in model (1) can be written as

$$Var_{\xi p}(\hat{\mathbf{B}}) = E_{\xi}[Var_{p}(\hat{\mathbf{B}})] + Var_{\xi}[E_{p}(\hat{\mathbf{B}})],$$
$$\approx Var_{p}(\hat{\mathbf{B}}) + Var_{\xi}(\mathbf{B}_{N})$$

where the approximations assume that the model expectation of the design variance is close to the finite population design variance, and $E_p(\hat{\mathbf{B}}) \approx \mathbf{B}_N$ (both of these assumptions hold at least approximately in a broad range of modeling and sampling contexts, but not in all). The above expression implies that the total variance of the design-weighted estimator $\hat{\mathbf{B}}$ is approximately equal to the sum of its design variance and the model variance of its finite population target $\mathbf{B}_N$. Both quantities can be estimated, the former based on the sampling distribution and the latter based on the superpopulation model. Typically, the former is inversely proportional to the number of primary sampling units, and the latter is inversely proportional to the population size. When the number of sampling units in the sample is much smaller than the population size, one can ignore the population variance component $Var_{\xi}(\mathbf{B}_N)$, and the total variance $Var_{\xi,p}\left(\hat{\mathbf{B}}\right)$ is then estimated by using design-appropriate methods for estimating the design variance $Var_p(\hat{\mathbf{B}})$. The above approach is quite general and has been described in the statistical literature for estimation of parameters in linear and nonlinear regression models, estimating equations, maximum likelihood, etc. The articles in Skinner et al. (1989) provide a good overview of design-based approaches for many of these models. Other cases not specifically covered by the above literature follow this same paradigm, as long as the concept of "equivalent finite population quantity associated with the model parameter" applies.

There are some important classes of statistical models for which design-based estimation will not provide valid estimators for the superpopulation model parameters. These include estimates of the variance components associated with random effects models, mixed effects models, structural equation models, and multilevel models. The fixed effects in these models can usually be estimated consistently, but not the variance components associated with the random effects, unless certain conditions on the sample sizes apply. If the random effects also need to be estimated, the design-based analysis becomes quite complex, and a model-based (unweighted) analysis might be preferable. Researchers working with ARMS data might also be interested in applying statistical methods that do not fall into the category of "model fitting," for example principal components analysis, clustering, classification. In those cases, the concept of "equivalent finite population quantity" does not apply directly, and it is sometimes unclear how to interpret the results of a weighted analysis. We discuss model-based inference for survey data in a subsequent section in this chapter.

In the cases for which a weighted analysis is possible, the advantages of the design-based approach are that it fully accounts for the design, and that it is robust to model misspecification, in the sense that it does not require the design to be ignorable with respect to inference for the superpopulation model. The disadvantages of the design-based approach are the lesser availability of software programs that can accommodate the design-weighted analysis, as well as the loss of efficiency relative to unweighted estimators, in cases in which the design is ignorable. The loss of efficiency is due to the fact that if the superpopulation model is correct and the design ignorable, then an unweighted estimator will typically have smaller variance than a weighted estimator. However, for large sample sizes in which the proportion of observations with very small weights is not overly large, the increase in the variance resulting from using the weighted estimate will not have a serious impact on the analytic conclusions, whereas using an unweighted estimate that can have serious biases could lead to misleading and inappropriate conclusions. In the following two sections, we discuss some design-based and model-based inference options that might be useful for researchers working with ARMS data.

Applying Design-Based Inference

The design-based paradigm described in the previous section should continue to represent a default approach for analyzing ARMS data, because it ensures that the estimates are free of bias due to the sampling process. For most users, the survey weights and the associated replication variance estimation weights make it possible to perform statistically valid model fitting and inference, without having to gain in-depth knowledge of the ARMS design and weight generation procedures. In this section, we address some additional issues related to implementation of the design-based paradigm related to ARMS.

In ARMS and most other large-scale surveys, the survey weights w_i that are included as part of the data set are not, strictly speaking, sampling weights, since that latter term is typically reserved for the inverses of the inclusion probabilities π_i. Survey weights are based on the sampling weights but are often adjusted for nonresponse and calibrated for known population quantities, either through modeling or poststratification. The effect of these adjustments for analytical studies is important but not often explicitly addressed in the survey estimation literature. We briefly discuss the current consensus on the effect of these adjustments.

Nonresponse adjustments to the weights are important to ensure that, as for the sampling design itself, the effects of the response mechanism are properly accounted for in estimation. The effect of ignoring the response mechanism in model fitting has been extensively studied, although usually from a model-based standpoint (see, e.g., Little and Rubin, 2002), and the principle of weighting by the inverse of the response probability is one of the possible solutions usually recommended in that literature. While weighting to adjust for both the sampling design and the response mechanism is therefore a valid approach, there is a critical difference between the sampling design and the response mechanism. The former is known by the survey statisticians responsible for creating the weights, while the latter is not and needs to be modeled. Hence, inference under joint sampling and response mechanism is often referred to as pseudorandomization (Oh and Scheuren, 1983), to differentiate it from pure (design-based) randomization inference. Sarndal and Swensson (1987) describe inference for finite population quantities under pseudorandomization, where the response mechanism is viewed as an additional phase of sampling.

For the purpose of estimating parameters of a superpopulation model, pseudorandomization would in principle need to be combined with the randomness of the model itself, using the same ideas as in the section on concepts for analytical inference for surveys. It is generally accepted that the survey weights represent the best available attempt by the statisticians responsible for a particular survey to account for both its design and the nonresponse. As long as the postulated response mechanism is correct, analyses performed using the joint sampling-nonresponse weights should result in consistent estimation of the finite population parameters, which are themselves model consistent for the superpopulation model parameters.

Calibration and the closely related model-assisted estimation approaches (Sarndal et al., 1992) are generally used to increase the efficiency of survey estimators and are also used for ARMS. While this increase in efficiency has been well documented for the case of estimating finite population quantities, the statistical literature on the effect of calibration on analytical inference for superpopulation model parameters is also very limited. Because properly applied calibration approaches create weights that remain "close to" the original inclusion probability weights and do not affect the design consistency of the weighted estimators, it is generally accepted that calibrated weights can be used in lieu of the original weights for the purpose of constructing design-consistent estimators for superpopulation model parameters. The recent monograph by Sarndal and Lundstrom (2005) describes estimation methods that account for both calibration and nonresponse, but it does not explicitly discuss analytical inference.

Variance estimators for survey-weighted estimators exist and are increasingly being implemented in a number of major statistical software programs. The most commonly used packages for design-based analysis are SUDAAN, STATA (survey module) and WesVar. Other packages incorporating some design-based methods are SAS, SPSS, and Mplus. However, for these latter programs, available procedures typically apply only to specific models and specific designs and currently almost completely ignore the effects of calibration and nonresponse adjustments. In situations in which the model and the design are covered by available software implementations, these offer a convenient way to perform design-based estimation and inference.

For these programs to correctly compute variances and perform statistical tests, using an analytic variance formula rather than using replication methods, it is necessary to specify detailed design characteristics, such as stratum identifiers and totals, cluster identifiers and weights. Hence, in order to be able to use this approach for ARMS, this information needs to be provided to users, which might cause confidentiality and disclosure issues if some information becomes too specific to be released as part of a data set for outside researchers. Because it might in principle be possible for ERS researchers to obtain more detailed design information from NASS, analyses performed by staff within the agency could be performed using existing statistical programs that implement design-based analysis and model fitting. This might require more access to confidential data than ERS currently has and would require ERS researchers, possibly in collaboration with NASS statisticians, to gain sufficient expertise in design-based methods to be able to evaluate the appropriateness of these programs for their modeling needs.

In order to accommodate a wide range of modeling needs and to fully account for postsampling weighting adjustments without having to release detailed design information, replication methods offer a useful alternative. A number of such methods are available, including balanced repeated replication, jackknife, and bootstrap methods (Wolter, 1985; Rao et al., 1992; Rust and Rao, 1996). In all these methods, each set of replicate weights undergoes the same range of weight adjustments as the original weights, so that they incorporate the adjustments in the variance estimation. The delete-a-group jackknife, the method currently implemented for ARMS, falls in this category and represents a valid design-based variance estimation method. Jackknife variance estimators make it possible to construct confidence intervals for parameters and perform Wald-type hypothesis testing on them. Both inference tools rely on the asymptotic normality of the estimator, which is generally reasonable for large samples.

NASS has made available some general-purpose programs for implementing the delete-a-group jackknife for estimating design-based variances. Some commercial software is also available for similar estimates. In some cases, these may be easier to learn to use than the specialized software written by NASS. Although there are a growing number of options to perform design-based variance estimation, in order to ensure consistency of the results across researchers for similar analyses, we suggest that researchers use the NASS-developed delete-a-group jackknife software whenever they want to use replication methods for their variance estimation. Although not always easy to apply, the NASS software has the advantage of being extremely flexible to apply while the jackknife is somewhat less dependent on assumptions. This is not to say that the jackknife is optimal in all situations, but the various estimators are relatively comparable in a wide variety of applications, and the jackknife is relatively robust to the circumstances of use.

A number of improvements to the current implementation of the delete-a-group jackknife would be very beneficial. A larger number of replicates should help alleviate the problem of small degrees of freedom encountered when multiple parameter estimates need to be tested jointly. More careful stabilization of the individual weights in the replicates is also advisable, to avoid either negative or unduly large weights that would result in unstable variance estimates. Alternative jackknife-based methods, such as the linearized jackknife (Yung and Rao, 1996; Binder et al., 2004), might provide a useful alternative methods for situations in which users are fitting models on small domains or for rare characteristics. Because of the importance of design-based estimation for many purposes, revision of the replicate generation process should be a high priority and implemented as soon as feasible as part of an overall strategy for improving data analysis.

Recommendation 7.3: NASS should investigate and implement improvements to the current jackknife replicates to make them more useful for the types of analyses performed by users in ERS and other organizations. In particular, NASS should increase the number of replicates and apply bounds to the magnitude of the weight adjustments.

Applying Model-Based Inference

As an alternative to the weighted, design-based analysis described above, it is also possible to consider a model-based analysis. This approach has a number of advantages, most importantly the fact that a wider range of statistical methods for analysis become available, many of which are already implemented in statistical software programs. The critical issue in this type of analysis is to ensure that the sampling design (and, as discussed above, the nonresponse mechanism) is ignorable with respect to the model of interest. As noted in Little (2004), "the major weakness of model-based inference is that if the model is seriously misspecified, it can yield inferences that are worse (and potentially much worse) than design-based inferences." Therefore, unless the researcher has been able to determine that, to the best of his or her knowledge, the design and nonresponse are either a priori ignorable or sufficiently accounted for in the model, the results from a model-based statistical analysis will not be scientifically credible. Model building, model estimation, and model checking for survey data are difficult and time-consuming tasks that should be attempted only by experienced researchers with knowledge in both survey statistics and statistical methods in the subject-matter discipline.

As noted earlier, valid model-based analysis of survey data requires a model for which the design and nonresponse are ignorable. In the regression context, this is typically achieved by expanding the model by including so-called design variables as predictors, by themselves and in interaction with other predictors. In the more general modeling context, incorporating the design and the nonresponse mechanism requires explicitly modeling what is seen as the important dimensions of selectivity (see below). In the case of multistage or clustered designs, it might also be necessary to specify a variance structure to account for design-induced correlation. Although the specific techniques for specifying the variance structure to account for design-induced correlation are not suggested here, they could include generalized estimating equation (GEE) methods, hierarchical modeling, or simply adding design variables to a regression model. In order to construct this expanded model, detailed information on the sampling design and the nonresponse characteristics of the survey is required, and each analysis may require different considerations for how to incorporate the informativeness of the sample in the model. It should be noted that the resulting model can be quite complex, both in terms of the mean structure and the variance structure.

The process of building a model for which the design and nonresponse are ignorable is often iterative. Once a suitable candidate model is specified, it is necessary to determine whether the design and the nonresponse are indeed ignorable. This can initially be done informally, through a comparison between the estimates obtained by traditional model-based methods and those obtained by a fully weighted analysis. If these estimates differ substantially, it is very likely that some aspect of the design is not yet fully incorporated into the model, so that further model expansion is required before proceeding with a model-based analysis. However, even when individual estimates are approximately similar, ignorability is not guaranteed and more formal statistical checks should be performed. In addition to general model diagnostics (e.g., residual plots, outlier detection) and goodness-of-fit tests on all aspects of the model, particular attention should be paid to possible effects related to the design variables, for example by plotting model residuals against survey weights. For a discussion of diagnostic plots for survey data, see Korn and Graubard (1999, Chapter 3.4).

A limited number of formal statistical tests are available to directly assess the effect of the design on regression model parameter estimates, based on the difference between the sample-weighted and the unweighted estimators (e.g., DuMouchel and Duncan, 1983; Fuller, 1984; Nordberg, 1989; and other references in Pfefferman, 1993, Sec. 4). The test procedure by DuMouchel and Duncan (1983) is particularly easy to implement, because it can often be performed as part of a fully model-based analysis.

Instead of incorporating the design variables as predictors in a regression, a more sophisticated approach involves modeling the sampling design and the nonresponse mechanism as part of the analytical model. This implies that the probability structure of the design becomes part of the model, and a full likelihood that includes the original model and the various selection mechanisms needs to be constructed. In situations involving complex stratification, clustering, and multiphase sampling, simply adding the design variables as predictors is often not sufficient, and this type of explicit modeling is required. Some relevant work in this area includes Chambers (1986), Skinner (1994), Breckling et al. (1994), and Pfeffermann and Sverchkov (1999).

As an alternative to the fully model-based approaches described above, some researchers use the design-based (weighted) estimator but evaluate it from a model-based perspective. This has the advantage that the estimator itself will be design consistent and hence avoids the main risk of model misspecification due to selectivity issues. The other advantage is that many software programs that do not do proper survey inference are still able to include weights in their estimation routines. However, using standard software to compute the weighted estimates of the unknown parameters does not generally provide either correct design-based or correct model-based confidence intervals or statistical tests of significance. The reason for this is that in most statistical programs, the weights are used as adjustments for heteroskedasticity, with the sampling weight taken as the inverse of the variance of the observation. Hence, unless the model is truly heteroskedastic with observation variances related to their inclusion probabilities, the variance estimator of the weighted estimator is incorrect. This is true regardless of whether the weights have been normalized or not (normalized weights mean that they have been scaled so that they sum to the sample size).

However, the idea of considering the model properties of the design-weighted estimator does have merit in assessing the ignorability of the design. If weights are treated as fixed constants rather than heteroskedasticity variance adjustments in the model fitting, then the resulting estimated variance of the model estimators will target $Var_{\xi}(\hat{\mathbf{B}})$. A few programs might have built-in options to treat the weights as constants, while for others, special macros or functions would have to written. In the specific case of a linear regression model with uncorrelated errors, some programs, including STATA, have implemented the Huber-White robust sandwich estimator, which is an appropriate estimator for $Var_{\xi}(\hat{\mathbf{B}})$. The model-based variance $Var_{p}(\hat{\mathbf{B}})$ is generally not equal to $Var_{p}(\hat{\mathbf{B}})$, which is the correct variance from a design-based perspective and an approximation to the full variance $Var_{\xi p}(\hat{\mathbf{B}})$, as explained in the section on concepts of analytical inference for surveys. However, when the design is ignorable with respect to the model being fitted, then it is usually the case that $Var_{p}(\hat{\mathbf{B}}) \approx Var_{\xi}(\hat{\mathbf{B}})$. This motivates an informal but useful way to assess the presence of a nonignorable design for the model-based analysis of a weighted estimator, by comparing the estimate of $Var_{\xi}(\hat{\mathbf{B}})$ and the estimate of $Var_{p}(\hat{\mathbf{B}})$ obtained by a design-based method, for example the ARMS jackknife weights. If these two estimates are very different, then the design is likely to be nonignorable and the model-based inference is not reliable.

While the model-based analysis of weighted estimators may seem attractive because of its simplicity, it is not robust to model failure. Some ERS researchers have used the Huber-White robust sandwich estimator available in the STATA package. This estimator is used by STATA whenever a probability weight is specified for all procedures that allow the "robust" option. It is also used as the default variance estimator for all weighted analyses in NLOGIT, and in SAS the MIXED and GLIMMIX procedures can compute the Huber-White estimator by specifying the EMPIRICAL option. However, such model-based robust estimators of the variance are still using some (weaker) model assumptions that may not necessarily hold. If they do hold, the model-based robust estimators would be similar to the design-based estimator, such as would be obtained from a delete-a-group jackknife estimator.

If, however, the assumed model is indeed true, then the weighted estimators of the parameters of interest are less efficient under the model than the unweighted estimators. Hence, researchers who are interested in conducting a fully model-based (unweighted) analysis should consider incorporating the design into the model as described above.

Recommendation 7.4: NASS and ERS should investigate the feasibility of providing sufficient information on the design and nonresponse characteristics of ARMS, in order to perform design-based statistical analysis without using the replicate weights and to allow users to incorporate design and nonresponse information in model-based analyses.

ERS researchers in particular might be interested in investigating the appropriateness of these model-based approaches for some types of analyses. Access to the full set of microdata as well as the element-level design information could put ERS in a unique position with respect to the ability to evaluate alternative modeling approaches for ARMS, obtaining the most statistically rigorous and efficient estimators for their models of interest. Because of the importance of fully accounting for the design and nonresponse in the model, this might require ERS researchers to gain fuller access to potentially confidential portions of the ARMS design data and develop a more in-depth understanding of survey design issues and statistical survey expertise.

Recommendation 7.5: ERS should build an enhanced level of in-house survey statistics expertise, in cooperation with NASS. The specialized expertise in both econometrics and survey statistics needed to accomplish this is currently not present in ERS and is likely to require a significant effort in recruiting and training.

GUIDE FOR RESEARCHERS

When researchers first started using ARMS data, the focus of the analysis was primarily on descriptive quantities of the population, such as means, proportions, and totals for the complete population or for subpopulations. As the richness of the data for more in-depth analysis became apparent, the demands for taking fuller advantage of such a rich data source grew. However, questions on the appropriateness of certain methods also became more apparent. As a result, researchers at ERS and elsewhere would benefit greatly from guidance on how to approach certain commonly used analytical methods. This would ensure that analyses done using ARMS data are conducted appropriately, make it easier for new users to begin working with those data, and provide a clear indication to the user community that NASS and ERS are committed to ensuring that this survey meets a high standard. It would also help officials at ERS explain the rationale behind the approaches adopted in the production of analytical outputs.

In view of this need, the ERS staff has taken the initiative to develop materials to assist outside users, albeit on an ad hoc basis. One such publication is Robert Dubman's ERS staff paper, "Variance Estimation With USDA's Farm Costs and Returns Surveys and Agricultural Resource Management Study" (Dubman, 2000). This useful paper was published with caveats. It was "reproduced for limited distribution to the research community outside the U.S. Department of Agriculture and does not reflect an official position of the Department." It serves a useful purpose in that it provides an overview of survey estimators, sample design, hypothesis testing, disclosure rules, and reliability measures for the two surveys and covers sums, ratios, means, multiple regression, binomial logit analysis, and order statistics. It is a beginning, in that it addresses one way to analyze the data, but the prescription does not cover all types of analysis and analysis at all scales, for example, small versus large samples.

In the panel's view, the Dubman paper is a good primer on how to use the existing weights and replication method, but it does not cover the full range of possible applications of the ARMS data. The user's guide should provide background on the issues associated with estimation for survey data (as we attempted to do, at a higher level of abstraction, in the current document) and also provide guidance for as close as possible to the full range of modeling and estimation problems encountered by users of the ARMS data. Because the range of estimation scenarios is so broad, it was not possible for us to recommend a single approach for all of them. Therefore, instead of attempting to recommend specific methods for all of them (or reverting to a one-size-fits-all approach), our goal was to provide a background on the issues and a set of recommendations on how NASS and ERS can broadly address the estimation needs of the ARMS data user community.

The panel recommends but has not developed a user's guide. This is because of our view that the process of developing a guide should be managed by the USDA and conducted in a truly collaborative environment. The guide would best be designed jointly by experts at ERS and NASS working with key data analysts.

To help set the stage for the necessary user's guide, this section builds on the previous discussion of the merits of adopting a design-based approach for many of the analyses. In this section, we now focus on some of the specific questions that arise from using the delete-a-group jackknife method to make inferences from the survey data.

To assist ERS in preparing the necessary user's guide, we give a specific example of the type of issue that should be included. Suppose one is interested in estimating a production function such as

$$Y = f(X_1,X_2;R) + \varepsilon,$$

where Y is output, the X_1,X_2 are inputs, R is a measure of the farm's environment or organization, and ε denotes residual stochastic noise. Initially, the researcher may specify a simple but restrictive functional form for a regression model:

$$\log(Y) = a + b_1 \log(X_1) + b_2 \log(X_2) + b_3R + \varepsilon.$$

This is a fairly standard representation of a production relationship. It is quite restrictive, and economic theory suggests that a more flexible form—allowing for a wider range of input substitution, scale relationships, and environmental interactions—might be called for, for example:

$$\log(Y) = a + b_1 \log(X_1) + b_2 \log(X_2) + b_3R + c_{11}[\log(X_1)]^2 + c_{22}[\log(X_2)]^2$$
$$d_{12} \log(X_1)\log(X_2) + d_{13} \log(X_1)R + d_{23} \log(X_2)R + \varepsilon.$$

This new model includes 15 parameters. To decide which model is appropriate, design-based methods can be used to apply the Wald test to compare the full and reduced models (on 5 degrees of freedom) for the nested hypothesis or an appropriate design-based F-test that accounts for the degrees of freedom associated with the estimate of the variance-covariance matrix. A weighted likelihood-ratio test could also be computed, although the distributional properties of such a test are less well known. It has been suggested that a Rao-Scott adjustment to the likelihood ratio could be applied, but, for most purposes, we would suggest using the Wald test. The guide could make recommendations to allow users to choose between these different methods.

Several other common questions face researchers using ARMS data, for which the guide would be extremely valuable. For example, one important question to address is the most appropriate approach (design-based versus model-based) for analysis for different types of questions. The best method for analysis depends on the survey design, the sample size, and the type and reason for the analysis. Simple linear models may need a different treatment than hierarchical models, small domains may need different procedures than large ones, etc. When the sample size is small, a model-based approach may be preferred to a design-based approach, especially if the design-based approach leads to much higher variances (see Kalton, 1983).

When a particular method is known *not* to be appropriate, then this should be explicitly mentioned in the guide. For example, design-based estimation of random effect variances in hierarchical models is problematic. Caveats should also be added regarding the use of such methods as clustering and classification for data coming from a complex survey such as ARMS. The limitations of the delete-a-group jackknife should also be explicitly addressed in the guide, including the limit on the number of parameters that can be tested simultaneously.

It should be noted that when the sampling is ignorable, the model-based approach to inference could lead to better inferences than the design-based approach; however, this approach may not be as robust to departures from the model, especially with respect to the correlation structure of the model errors. A discussion of the pitfalls in modeling under nonignorable sampling, as well as guidelines on model construction and on testing for nonignorability of the design, would be very useful.

The guide should also discuss the issue of using weights in standard nonsurvey software, which provides the correct point estimates but potentially misleading inference. It might be possible to determine that certain programs provide inferences in this manner that are close to correct (possibly after weight normalization), in which case these could be recommended. For example, the Huber-White robust sandwich estimator for linear regression provides a way to estimate the model-based variance of the weighted estimator. For other programs, no automatic way is available to obtain valid variances for the weighted estimators, so that additional programming may be required. In general, the guide should stress that the model-based variance needs to be compared with the design-based variance, with any significant deviations between the two a cause for concern. The discussion in previous sections makes additional points that could be included on these topics.

Many analyses are based on domains (subpopulations) of the entire national population, so questions on the appropriateness of the delete-a-group jackknife come into play. Analysts notice that the sum of the domain weights for each jackknife replicate varies greatly among replicates, and they wonder if the variance is unduly inflated as a result. A standard design-based method is still appropriate here, since the fact that the sum of the weights varies appreciably is a reflection of the randomness in the sample for estimating domains. For such quantities as domain means or regression coefficients, the effect of this variability may be less pronounced, compared with the estimation of domain totals. However, compared with a model-based analysis, the variances may appear to be large. This demonstrates more the deficiency in the model-based approach, which does not account for the randomness in the domain sample size, rather than reflecting badly on the design-based approach. This should be explained in the guide, so that researchers encountering this problem will understand the reason.

Another question of interest to researchers is how to integrate data from more than one survey. These questions are more complex, especially if there is a need to use a coordinated jackknife, so some guidelines to users would be very valuable. In this area ERS and NASS may not currently have the necessary expertise. In general, there are two choices for analyzing the data when it is thought that the parameters of interest are similar across surveys: (1) the separate approach, in which the estimates are produced for each survey separately and then combined using a composite weight, and (2) the combined approach, in which the files are combined and analyzed as one file with a possibly adjusted weight variable. Under the separate approach, it is possible to use meta-analytic methods to combine evidence from different surveys (e.g., Zieschang, 1990). If the combined approach is pursued, it is always recommended that the assumptions about equality of the parameters of interest across surveys be subjected to a statistical hypothesis test.

Finally, the guide should also include a list of relevant references for researchers dealing with survey data. Some possible references are Lohr (1999), Pfeffermann (1993), and Korn and Graubard (1999). A gentle introduction to the basic issues of design-based and model-based inference for survey data is provided by Carrington et al. (2000), which is available online at http://www.ces.census.gov/index.php/ces/1.00/cespapers.

Recommendation 7.6: ERS and NASS should collaborate on writing a Guide for Researchers for performing multivariable analyses using data from complex surveys, particularly data from ARMS. In areas in which expertise is not available for writing parts of such a guide, expertise should be sought from the statistics and economics community, especially those with experience in the analysis of survey data from complex survey designs.

8

Dissemination

In reviewing the dissemination procedures in place for the Agricultural Resource Management Survey (ARMS), the panel drew on our own experience as users of the aggregated and microdata and reached out to data users in government and the research community. This chapter outlines the steps the Economic Research Service (ERS) and the National Agricultural Statistics Service (NASS) have taken to improve access. Internet capabilities have grown with the ARMS briefing room and the various levels of access and resources available.

Both ERS and NASS have taken some significant steps in the past few years to increase access to ARMS data, in both aggregated and individual record formats. Although hard copy publications are still prepared and offered, aggregated data are provided via the Internet for the most part, and individual record data are accessed mainly by computer terminals at ERS or NASS state offices.

Hard copy publications are fast becoming a thing of the past. NASS publishes two hard copy reports from ARMS. The first, called *Agricultural Chemical Usage—Field Crops*, is released in May following the Phase II data collection. The second report, *Farm Production Expenditures*, is compiled from the Phase III and released in July. ERS prepares several hard copy state, regional, and national reports using ARMS data, including *Commodity Production Costs and Returns, Farm Operating and Financial Characteristics* and the *Annual Report to Congress on the Status of Family Farms* (National Agricultural Statistics Service, 2007c).

As a general rule, however, hard copy sources are being supplanted by websites supported by sophisticated web tools, microdata files, and special centers for access to confidential data in a protected environment. These new forms of dissemination have increased the need for training, formal mechanisms for communication with users, and tightened procedures for control over the data.

NEW DATA SOURCES AND PRODUCTS

The most ambitious new product is the ERS-managed online ARMS briefing room.[1] This website provides summary data as well as extensive ARMS documentation and access to questionnaires. The data available on the website are retrievable in the form of tailored reports and summary tables. The tables provide means and standard error estimates, they can be saved as comma separated value (CSV) or Excel files, and they have a capacity for graphical display of data.

There are several means of accessing less aggregated data. The ERS produces special tabulations, typically for government agencies. ERS has provided data and research support using ARMS directly to the U.S. Department of Agriculture (USDA) Office of the Secretary, the Office of the Chief Economist, Research, Education and Economics, the National Resource Conservation Service (within NASS), the Farm Service Agency, the Foreign Agricultural Service, the Animal and Plant Health Inspection Service, and the Risk Management Agency. ERS has also used the data to respond to requests from Congress, the Office of Management and Budget, the Bureau of Labor Statistics, universities, and international organizations.

The myriad uses of ARMS data require sophisticated access to data in order to depict increasingly complicated relationships. For example, recent analyses of the financial status of farming used ARMS data to prepare outlook information for farms and farm households by type of farm, size of farm, and U.S. region; balance sheets for farms by size of farming operation; a cumulative distribution of farms by economic cost-to-output ratios; and the distribution of key commodity production by production cost level. Other examples abound. ARMS data were used to provide a wide variety of information types:

- Distribution of government payments by farm size
- Farm household income by farm typology
- Impacts of energy price increases on farm businesses
- Impacts of seed price increases
- Characteristics of U.S. production of biotechnology-derived crops
- Crop insurance usage by typology, commodities, and regions
- Types of farm management practices

---

[1]Available: http://www.ers.usda.gov/Briefing/ARMS/

- Agricultural land ownership by women
- Cost of production for corn and soybeans
- Limited resource farmers by state
- Changes in American farming from 1988 to 2001

In addition to supporting research that depends primarily on ARMS data, ARMS contributes to other research, analyses, and situation-and-outlook work in federal and state governments, academic institutions and other organizations because it provides the basic cost-of-production and supply response information on which other analyses depend. Examples of ARMS use in such research include

- *USDA agricultural baseline projections to 2010.* Baseline reports provide long-run (10-year) baseline projections for the agricultural sector. The reports contain cost-of-production data that are the baseline for projections covering supply and demand for agricultural commodities, agricultural trade, and aggregate indicators of the sector, such as farm income and food prices.
- *Cotton.* Background and issues for farm legislation. Recently issued commodity background reports, developed to inform the farm bill debate, provide charts and discussion on the distribution of costs across farms and other data derived from ARMS.
- *Managing risk in farming.* Concepts, research, and analysis. This report on the risks confronted by grain and cotton farmers and risk management tools and strategies used at the farm level uses ARMS-based data on farmers' assessments of the risks they face and their use of alternative risk management strategies.
- *Effects of federal tax policy on agriculture.* Comparisons between financial and tax variables reported on farmers' tax returns and the ARMS data highlight differences between accounting for tax purposes and the underlying economic realities.

ARMS Web Tool

ERS advertises that any user can get customized data summaries provided by the publicly available online data tool at http://www.ers.usda.gov/Data/ARMS/. Tailored reports enable custom queries, in which users can select from a set of variables and customize the estimates they receive, refine queries with specific samples or populations, group summary statistics for comparisons, and choose among output options for results (tables, charts, etc.).

The basic customized data summaries available through the ARMS briefing room are broken into four major data topics: (1) farm structure and finance, (2) crop production practices, (3) commodity production costs and returns, and (4) featured states. Within each topic area users can create tailored reports from the survey data.

Farm structure and finance reports contain information on the structure and financial status and performance of U.S. farm operators, their households, and farm businesses. Crop production practices summaries include information about the status and trends in crop production practices for several field crops. Commodity production costs and returns summaries include statistics on the annual production costs and returns for major field crop and livestock commodities. Special reports for the 15-state oversample are available for the time period 2003-2005.

The ARMS web tool for customized summaries is a convenient and effective way to disseminate findings from ARMS to the general population. It is user-friendly, and the summaries and findings are easy to interpret. A major strength is that the ARMS web tool can also provide summaries of variables by categories or for multiple years. Individuals interested in ARMS can conveniently begin their exploration using the web tool. In the panel's view, ERS should continue to update and develop the web tool as their primary means of disseminating ARMS findings to the public and to data users. One suggestion for improvement is to continue to expand the list of variables that are summarized.

ARMS Extranet Web Tool

In addition to the publicly available ARMS web tool, ERS has developed an advanced statistical analysis web tool, which is available to approved researchers with proper authorization. Currently this web tool can perform regression analysis with variables and by categories that are predefined and provided by ERS. All output is screened to ensure data confidentiality.

The ARMS extranet web tool has great potential to facilitate researchers in the conduct of preliminary or full analyses of their models. It allows researchers more flexibility to estimate advanced models and the convenience of using the ARMS data remotely, without compromising data confidentiality. This tool has great potential to increase the availability of the ARMS data to researchers for whom it is costly or inconvenient to access the data on-site. Although it is useful for preliminary analyses with the data, the current functionality of the tool is limited to ability to support ordinary least square regressions with predefined variables, and this has prevented data users from fully using the tool for research. It can be made more useful for research purposes. One improvement would be to allow researchers to use, redefine, and combine variables in the raw data set instead of only the aggregate, predefined variables that are available now. A second improvement would be to continue to expand the models that are available for estimation (limited dependent variable models, etc.). Another improvement would be to allow for different years of data to be included in one model.

Recommendation 8.1: ERS should continue to improve the ARMS web tool by providing summaries on more variables and more subsets from ARMS, and to improve the ARMS extranet web tool by adding the ability to link over years and to more sophisticated models.

Special Tabulations

On occasion, users may request special tabulations of ARMS data beyond what is available in published or downloadable reports and tabulations. Special tabulations require a commitment of time and expertise of agency staff. Under ERS policy, the agency may provide special tabulations, reimbursed on the basis of staff time required to prepare the tabulations, provided the staff are available. ERS policy is that special tabulations, once prepared, are available to the public, and the agency reserves the right to disseminate the results of special tabulations on the agency website or in agency publications. NASS has a similar program (National Agricultural Statistics Service, 2007d) and has produced special requests for users for a fee—an example that is cited is a report for John Deere. The reports produced in this manner are nonproprietary and are published on the USDA website.

MICRODATA ACCESS FOR DATA USERS

The individual records derived from ARMS have been collected and acquired exclusively for statistical purposes under a strict pledge of confidentiality. However, with the recognition that the individual records are valuable in understanding the status of farms and farm families, procedures have been established to provide access to microdata in an environment that ensures that the confidentiality of the records is maintained.

Currently, approved data users can access the ARMS microdata either onsite at ERS or NASS offices. They can also access the data in a more limited form and run simple regressions online. As discussed in this section, alternative means of dissemination, particularly through the Census Bureau Data Centers, may increase access for some researchers or facilitate analysis that combines data sets.

Procedure for Accessing the ARMS Microdata

ARMS individual-record microdata can be made available to researchers and other government agencies that have collaborative projects with ERS or NASS that contribute to USDA's public-sector mission. These projects must be formally administered through a cooperative research relationship between ERS and a responsible research organization. Users apply by submitting a memorandum of understanding for research purposes between ERS and the research institution, an approved research project agreement, and a NASS confidentiality agreement. Users also must participate in a security briefing on the security and confidentiality requirements of using ARMS data. Once formal agreements have been signed, access to ARMS data is provided at ERS or NASS headquarters or at a NASS state statistical office, depending on availability of office and computer resources.

ERS takes these steps to ensure that ARMS data are used for statistical purposes only, not for other purposes, such as regulatory or enforcement purposes or release under the Freedom of Information Act. The objective is that confidential data will not be disclosed. In addition, all reports, publications, and releases based on ARMS data must pass strict nondisclosure reviews. Entities and individuals outside USDA with access to confidential survey data are subject to the same federal statutes that apply to USDA and its employees. Under these statutes, individuals who unlawfully disclose confidential data are subject to fines and other penalties. The procedures for accessing the microdata are designed to establish widespread and uniform confidentiality protections that cooperators must adhere to. They take advantage of the Confidential Information Protection and Statistical Efficiency Act of 2002, which grants "licensed agents" use of confidential data for statistical purposes, providing criminal and civil penalties for misuse.

Current Users and Projects

Researchers can access the data through their institutions or directly with USDA. Over 25 institutions, primarily academic institutions and some government agencies, have agreements in place to gain access to the ARMS data on-site at ERS or NASS offices or online via the ARMS extranet tool.

A number of ongoing initiatives are aimed at improving the dissemination of and access to ARMS data. They include cooperative agreements with the University of Illinois and the University of Minnesota, web design changes, and continuing research on alternative approaches to access.

On-site Data Access at ERS and NASS State Offices

To access the ARMS microdata at ERS or NASS state offices, researchers contact the NASS office in their state and arrange visits to estimate their models with the ARMS microdata. NASS offices provide data users with a secure computer on which the ARMS data and the statistical analysis system (SAS) software and Microsoft office programs are installed. Because time spent at NASS offices is limited by travel costs and work schedules, researchers need to arrive well prepared. They typically bring computer programs to the NASS offices and spend the majority of their time running and modifying their programs and compiling the results. Before they leave, ERS or NASS staff inspect all output and results for compliance with the confidentiality of the records.

For most university researchers, the availability of the ARMS data at the NASS state offices significantly reduces their travel costs in comparison to the alterative of traveling to Washington, DC. Nonetheless, many users find it inconvenient having to travel sometimes considerable distances to a state office. In addition, some have encountered difficulties in processing the data at the state centers. Users have called for making the data available in the state centers in a format that would facilitate the use of standard computer languages other than SAS, such as STATA, Limdeb, and Matlab. Policies on software vary from state office to state office. The software available may differ, and some state offices seem to be uncomfortable with allowing users to upload their own software to computers in the state office.

At the panel's session for ARMS data users at the 2006 meeting of the American Agricultural Economics Association (AAEA), the group consisted mainly of academic researchers but also included some private-sector users. They raised several concerns and made a number of useful observations about access to the ARMS data:

- NASS state offices vary with respect to computer and staff resources dedicated to ARMS research and the ability to host data users. Most are able to accommodate the specific needs of researchers with respect to their preferred time for visits or software needs, but some state offices require advance planning and scheduling of visits and are not very responsive during peak work activity periods.
- A suggestion for improving access was that ERS should create more comprehensive documentation of the content of the data files, so that researchers could better prepare before arriving at the state data center.
- Despite the relative convenience of the state centers, some researchers prefer to travel to Washington, DC, to access and work with the ARMS data at the ERS facility. This affords them special attention from knowledgeable ERS staff.
- Nonacademic and nongovernment researchers face particular difficulties in accessing ARMS data. Independent researchers cannot disaggregate the online data: they do not have access to the restricted data, and they often do not have the resources to request ERS staff to run a study for their purposes.

Census Bureau Research Data Center Program

An alternative means of allowing researchers to access federal government survey micro records is offered by the Census Bureau through its research data centers (RDCs). Incorporating access to ARMS microdata in the RDCs would expand the number of points of access, and it could facilitate the linkage of ARMS data with other data sets, which could add value to the models that are built.

The Census Bureau operates the RDCs under the authority of the secretary of commerce to grant *special sworn status*, which allows individuals outside the Census Bureau access for tasks that benefit the Census Bureau's mandated programs. This "benefit" test means that users must go through a number of steps, including working with an RDC administrator, and submit an online proposal, which is reviewed on scientific merit, benefit to the Census Bureau, disclosure risk, policy sensitivities, feasibility, and proof of a clear need for nonpublic data. These proposals are reviewed by both Census Bureau and external (academic) reviewers. If another agency developed or sponsored the data, necessary proposals are also required to be forwarded to that agency for review. If the proposal is approved after these steps, all researchers on a project must undergo a background check, provide a Social Security number, sign an oath that they will preserve confidentiality, and take data and information technology security training.

A recent decision by the Census Bureau leadership to consider allowing non–Census Bureau data at RDCs has made them a potentially attractive alternative to other agencies that now maintain their own data access programs. At the time this report was being prepared, the Census Bureau had begun to house confidential data at the RDCs for the National Center for Health Statistics and to work out arrangements for others to house their confidential data at the RDCs. An agency owning data housed in an RDC would be responsible for proposal review and setting disclosure standards but would use the RDC infrastructure, including online proposal capture, computer labs, and management capacity at the data centers. In return, the agencies would contribute to the cost of maintaining, upgrading, and expanding the RDC infrastructure.

Recommendation 8.2: USDA should consider extending the availability of ARMS microdata through the Census Bureau research data centers to increase access opportunities for using additional data sets and enabling researchers to match ARMS files with other data sets.

TRAINING FOR DATA USERS

ERS provides data users with information, documentation, and training through the Internet and a help desk operation. In the online ARMS briefing room, interested users can learn about the survey and how to apply for access to the microdata. Users can use the ARMS web tool to generate custom data summaries and obtain downloadable copies of all survey questionnaires. Users seeking access to the microdata are directed to contact the ERS Survey and Data Coordinator who provides help desk assistance, including relevant questionnaires, electronic copies of respondent booklets, NASS and ERS variable listings, format listings, summary programs, and artificially generated nonsurvey data for use in optimizing computer programs before visiting the state office or the ERS office in Washington, DC.

ERS provides initial statistical guidance to microdata users. Researchers are assumed to have adequate statistical and economic modeling skills to accomplish their projects, including sufficient SAS abilities. ERS sends copies of reports on variance estimation under the complex sample design in ARMS and offers samples of SAS programs that illustrate jackknife techniques and regressions. ERS also provides on-site advice and support to microdata users who wish to come to ERS for a few days of hands-on learning and experience.

ERS continues to explore ways to enhance access to ARMS data and training for users. A recent cooperative agreement between ERS and the National Opinion Research Center seeks to develop training guidelines for users on data security and on ways to use ARMS. ERS also organized a preconference workshop as part of the 2006 AAEA annual meeting entitled "The State-of-the Art in the Analysis of Survey Data with Complex Sample Designs." The objective of the workshop was to illuminate the underlying statistical issues in analyzing data with complex sample designs.

Many university researchers are interested in using the ARMS data for research but have not done so because of the steep learning curve and the travel costs associated with accessing the microdata. Training programs targeted at new data users and online tools for preliminary analyses are likely to be effective in attracting more data users. Because of the complexity of the survey design and the generally used estimation procedures, new data users will benefit from additional information and training to reduce startup costs of using the ARMS data.

At the panel's session for data users at the 2006 AAEA meeting, discussions among the existing data users revealed that the documentation related to ARMS is generally helpful and adequate for most purposes. However, it also became clear that many data users are not fully aware of all available documentation and resources, mostly because the materials are delivered on a case-by-case basis and because researchers visit NASS state offices without direct contact with other data users or ERS staff. Therefore, it would be helpful for the existing information to be packaged into a data user manual and delivered through training programs.

The panel thinks that the complexity of ARMS access requirements and the data themselves, combined with the high costs and steep learning curve involved, especially for first-time users, have seriously impaired the value of ARMS as a research tool.

Recommendation 8.3: ERS should provide more training for new data users, including developing a data user manual, which also includes the recommended guide on statistical estimation (see Recommendation 7.5), and offering training workshops.

USER FORUMS AND FEEDBACK

In the ARMS briefing room, users can submit an open-ended feedback form to the ARMS team. The site also includes a few quick links through which users can discuss specific issues, such as ARMS content issues, and can request special tabulations or access to the online tools or the microdata. Recommendations for specific content to include in the survey and requests for special tabulations go directly to the data and survey coordinator in ERS. Users can also sign up to receive the ARMS update newsletter, which is also available through the briefing room.

Since the feedback options available through the briefing room go directly to ERS staff, they do not allow users to actively discuss issues with one another. At the panel's data user session at the AAEA's 2006 annual meeting, users and potential users of ARMS data expressed interest in developing a process to facilitate communication within the user community as well as an ongoing feedback process to NASS and ERS. One idea is to schedule regular data-user meetings, in which researchers exchange ideas among themselves and offer feedback to ERS. Another idea is for ERS to help launch a "wiki" or some other type of interactive online documentation for ARMS, in which researchers themselves post ARMS-related information. This would also allow users to share programs, tips, and ideas about the ARMS data.

DATA CONSISTENCY

Although the panel has found no reason to doubt that ARMS data are equal in quality to those of other federal statistical agencies, occasionally clear errors have been found in the data by users, despite the review at NASS, ERS, and the state offices. A formal revision policy would encourage users to report outliers and data inconsistencies. As data files are updated, it will be useful to keep records of data corrections and updates.

In order to access and process summary and microdata in electronic form, users must use codes that identify individual variables, and these codes change over time (see Table 8-1). Changes in questions over time are also another serious complication for users. A problem many of them face is that the questionnaires change over time and the questions change between versions—an issue discussed in this report. Table 8-1 provides examples of codes for the variables included in the debt section of the survey.

For these variables, there is a high level of consistency in the questions included in the survey every year, but the codes change and sometimes change back. NASS and ERS should develop a variable naming convention that does not permit code changes unless underlying concepts change. Although any arbitrary set of codes could be used by NASS and ERS in developing the data, but these should be mapped into a stable system of nomenclature. NASS and ERS have a very strong comparative advantage in judging the comparability of variables over time, particularly since they need such information to perform their own work.

Like the variables themselves, the summary statistics are not consistent over the years. ERS should make the names and definitions of the summary statistics as consistent over the years as possible. This can be solved for future years by a policy of maintaining consistency in the names and definitions of summary statistics. In addition, ERS would be providing a

TABLE 8-1 Examples of Codes for Debt Variables, 1996-2005

Year	Lender Type	Loan Balance	Interest Rate	Loan Term	Year Obtained
1996	R611	R615	R619	R623	R627
1997	R938	R939	R940	R941	R942
1998	R931	R932	R933	R934	R935
1999	R1001	R1002	R1003	R1004	R1005
2000	R1001	R1002	R1003	R1004	R1005
2001	R1001	R1002	R1003	R1004	R1005
2002	R1161	R1162	R1163	R1164	R1165
2003	R1001	R1002	R1003		R1005
2004	R1001	R1002	R1003		R1005
2005	R1001	R1002	R1003	R1008	R1005

service to its users by revisiting the definition and naming of summary statistics from previous years to make them consistent with the names used in the most recent years. Only in this manner can a consistent time series for the summary series be ensured. In addition, if possible, the introduction of new aggregate or categorical variables (such as the new "combined farm" typology) should be accompanied by the revision of older data sets to include those variables. At a minimum, ERS should, when possible, provide computer code to users that provides for an intertemporally consistent definition of the key variables they publish.

Recommendation 8.4: Database management practices should include a system for managing and reporting errors found by users for consistent labeling of the codes for raw variables, and for using the consistent names of the ERS-created summary variables over time.

TABLE 8-1 Continued

Percent for Farm Use	Loan Guarantee	Type of Loan	Purpose of Loan	Number of Other Loans	Balance on Other Loans
R631	R639	R635		R643	R644
R943	R944			R966	R967
R936	R937			R959	R960
R1006	R1007				
R1006	R1007			R1036	R1037
R1006			R1007	R1036	R1037
R1166			R1167	R1196	R1197
R1006	R1008 R1009	R1004	R1007	R1046	R1047
R1006	R1008 R1009	R1004	R1007	R1046	R1047
R1006		R1004	R1007	R1046	R1047

9

Conclusions and Recommendations

The Agricultural Resources Management Survey (ARMS) is a complex undertaking. It began, just over a decade ago, as a melding of data collected from the field, the enterprise, the farm, and the household, in a multiphase, multiframe, and multiple mode survey design, and it has increased in complexity in the ensuing years as more sophisticated demands for its outputs have expanded. Today, at the outset of its second decade of operations, the survey faces difficult choices and challenges, including a need for a thorough review of its methods, practices, and procedures. This report contributes to that necessary endeavor.

This chapter summarizes the panel's review in a short statement of our major conclusions and presents the recommendations that appear in Chapters 2 through 8.

CONCLUSIONS

ARMS is an invaluable source of information on the current state of American agriculture, as well as the sole source of some important information on the linkages between fields, farms, and families that serves to illuminate the challenges faced by agriculture policy makers and farm families. Because the survey is so critical to understanding agriculture, it carries a special burden. Its methods, practices, and procedures must be designed to yield data of impeccable quality in view of their uses, and the data must be made available to the research community both inside and outside the federal government in order to generate the improved analytical knowledge the data makes possible.

The panel's review of several aspects of the survey's statistical quality was challenging. At several points in this report, some of the methods and practices used in ARMS are characterized as "unique" or "unconventional." In large part, the unique nature of the survey is due to it complexity, with multiple modes and phases and with a goal to collect, classify, and aggregate several types of information from three interrelated but not entirely overlapping reporting units. ARMS reflects some unique practices that are part of the U.S. Department of Agriculture's (USDA) way of doing business, such as its board review process, which are not within the panel's purview to assess. Nonetheless, the panel has been able to document and assess the adequacy of the survey design, data collection, analysis, and dissemination.

The panel concludes that appropriate attention is being paid by the National Agricultural Statistics Service (NASS) and the Economic Research Service (ERS) to the basic elements of survey quality, although much more could be done to improve important features of the survey. Several aspects of survey operations need more attention, including the employment of analytical tools to investigate the quality of survey responses, additional control and further automation of the interview process, shifting the focus from nonresponse rates to nonresponse bias, introduction of new methods of imputation of missing values and documentation of the results of imputation, improvements to variance estimation that are more compatible with the types of data analysis uses that are now employed, and more attention to facilitating access to the data files for research and analysis.

In addition to identifying areas of needed improvement in current methods and practices, our review identifies several emerging challenges. These challenges are associated with the changing structure of farming, overall trends in federal surveys—such as the growing difficulty of obtaining satisfactory survey response—and the growing sophistication of survey data users, both inside and outside the federal government. The agencies have attempted to respond to these challenges with some foresight—adding new questions, testing such initiatives as incentives for increasing reporting, developing proposals to collect longitudinal data, and enhancing the provision of microdata files in a protected environment. Our review leads to the conclusion that several areas still need attention, and the recommendations that follow may be considered a roadmap to the future for ARMS.

We are aware that our list of recommendations is long and that some of them will be costly to implement. Full implementation of all of them would require a significant fraction of the ARMS budget. In our view, if additional funds cannot be obtained, at least those recommendations involving methodological research and development directly related to data quality assurance should be undertaken, even at the expense of reducing the size or scope of the survey. For other costly recommendations, notably the training programs and other services for data users outside USDA, additional funding could reasonably be sought, even from unconventional sources in the user community. For example, the land grant universities could be asked to support, and perhaps to assist in implementing, the training and data access improvements. The universities rely on other sources of USDA funding, through the Cooperative State Research, Education, and Extension Service, which might be interested in funding competitive National Research Initiative or other grants for these purposes.

RECOMMENDATIONS

Although the panel did not attempt to prioritize the issues or our recommendations, we do draw special attention to the need for an ongoing, joint, and appropriately funded methodology research and development program. Such a program needs adequate resources both to support current and future research, development, and statistical analysis needs throughout the implementation of the ARMS and to assess and manage the quality of the data. We also call attention to the need for better channels of communication with providers and users of the data. These initiatives will require an infusion of funding and, in the case of ERS, enhancement of staff expertise in mathematical statistics and data analysis skills. The panel believes that all of the recommendations are feasible and important, but some are more important than others and are worded to convey that immediacy.

In the pages that follow, we present the recommendations that appear in context throughout the report.

Data Integration and Relevance

Recommendation 2.1: The National Resources Conservation Service, NASS, and ERS should engage in a focused research and testing program and use experience with integrating the Conservation Effects Assessment Project and ARMS to assess the feasibility of integrating ARMS with other surveys and data sources.

Recommendation 2.2: In preparation for funds becoming available for a longitudinal design of ARMS, ERS and NASS should systematically conduct research and explore the need for and feasibility of obtaining panel data from ARMS. Furthermore, as a test of the power of such information, more use should be made of the existing longitudinal microdata from the Census of Agriculture. One possible approach would be to create a pseudo-panel of such data. Another would be to make a retrospective link between the Census of Agriculture and a year of ARMS.

Survey Management

Recommendation 3.1: The ARMS program should have structured mechanisms in place for stakeholder feedback and discussion on ARMS, beyond what is currently done, such as organized stakeholder forums, with some obligation to respond. Specifically, USDA should solicit input in developing the survey from stakeholders from within USDA and from other government agencies, universities, professional associations, and the private sector.

Recommendation 3.2: The NASS Advisory Committee on Agriculture Statistics should expand its scope to include an annual review of ARMS.

Recommendation 3.3: ERS and NASS should establish an ongoing, jointly sponsored, and appropriately funded methodology research and development program. Such a program should provide adequate resources to support current and future research, development, and statistical analysis needs throughout the implementation of ARMS and to assess and manage the quality of the data. If new funds cannot be obtained, funds from existing programs must be reallocated.

Recommendation 3.4: NASS and ERS should commit resources to developing a five-year plan tied to the Census of Agriculture for ARMS content, coverage, and methodology. The agencies should develop measures to control changes during the five-year period to minimize disruptions to the time series of the core content in ARMS.

Sample and Questionnaire Design

Recommendation 4.1: The methodology research and development program the panel recommends should systematically (1) evaluate current instruments and practices, (2) collect data that inform both the revision of existing items as well as the creation of new items, (3) test revised instruments before they are put into production, (4) use experimental control groups to evaluate the differences between the old and new questionnaires, (5) improve understanding of respondent record-keeping practices and their effect on survey quality, and (6) designate a subsample of the existing ARMS sample for research and testing purposes. Key parts of this work would best be conducted in a cognitive or usability laboratory facility. It would be enabled by obtaining a generic clearance from the Office of Management and Budget for testing of all phases of the survey to allow for broader cognitive testing, evaluate the quality of data reported in response to each question, and evaluate the impact of mode of data collection across the three phases.

Recommendation 4.2: ERS and NASS should improve the consistency of variables across ARMS versions and over time.

Recommendation 4.3: NASS and ERS should explore the collection of auxiliary information on a formal basis, as well as feasibility of enriching the ARMS data files with information from administrative data sources, geospatial data, and the like.

Data Collection

Recommendation 5.1: ARMS should use automated means to collect paradata on interviewer assignments to cases, the relationship between the interviewer and the sample farm respondent (i.e., whether they know each other), demographic characteristics of the interviewer and the characteristics of the sample farms for nonrespondents that are coordinated with information obtained for respondents, either through the interview or interviewer observation. These paradata could be used to determine the need for additional research on the impact of the relationship between the interviewer and the respondent on the quality of answers. This data collection can best be facilitated using computer-assisted technologies.

Recommendation 5.2: NASS should systematically explore the consequences of interviewer departures from standardization in the interview. To facilitate this, NASS should collect paradata on the frequency with which interviewers follow the order of the questionnaire, read questions as worded, provide clarification, and similar indications of departures from standardized procedures.

Recommendation 5.3: NASS should use available analytic tools, for example, cognitive interviews, interviewer debriefing, recording and coding of interviews, and reinterviews, to investigate the quality of survey responses.

Recommendation 5.4: NASS should move to computer-assisted interview and possibly web-based data collection, after research and testing to determine possible effects of the collection mode on the data. Computer-assisted personal interviews and web-based data collection will provide opportunities to increase timeliness, improve data quality, reduce cost, and obtain important paradata.

Recommendation 5.5: NASS and ERS should develop a program to define metadata and paradata for ARMS so that both can be used to identify measurement errors, facilitate analysis of data, and provide a basis for improvements to ARMS as part of the broader research and development program the panel recommends.

Nonresponse, Imputation, and Estimation

Recommendation 6.1: NASS should routinely report ARMS case dispositions linked across survey phases to provide the foundation for appropriate response rate calculations for Phases II and III.

Recommendation 6.2: All published ARMS response rates for Phase II and III should be calculated to reflect the nonresponse from the preceding phase(s).

Recommendation 6.3: The nature of the ARMS nonresponse bias should be a key focus of the research and development program the panel recommends. This research and development program should focus, initially, on understanding the characteristics of nonrespondents.

Recommendation 6.4: The research and development program should continue NASS's work on both public relations and incentives, and it should do so with a focus on nonresponse bias, not simply nonresponse rate.

Recommendation 6.5: The research and development program should analyze whether there are differences in ARMS unit and item nonresponse rates between census and noncensus years, with an eye toward deciding whether making ARMS mandatory would improve data quality.

Recommendation 6.6: The research and development program should examine how questionnaire design and interviewing changes could reduce item nonresponse.

Recommendation 6.7: NASS and ERS should consider approaches for imputation of missing data that would be appropriate when analyzing the data using multivariate models. Methods for accounting for the variability due to using imputed values should be investigated. Such methods would depend on the imputation approach adopted.

Recommendation 6.8: All missing data that are imputed at any stage in the survey should be flagged as such on files to be used for analysis.

Recommendation 6.9: NASS and ERS should provide more clarification and transparency of the estimation process, specifically the effect of calibrations on the assignment of weights and the resulting estimates.

Methods of Analysis

Recommendation 7.1: NASS should continue to provide sampling weights with the ARMS data set, combined with replication weights for variance estimation.

Recommendation 7.2: NASS and ERS should continue to recommend the design-weighted approach as appropriate for many of the analyses for users of ARMS data and as the best approach for univariate or descriptive statistics.

Recommendation 7.3: NASS should investigate and implement improvements to the current jackknife replicates to make them more useful for the types of analyses performed by users in ERS and other organizations. In particular, NASS should increase the number of replicates and apply bounds to the magnitude of the weight adjustments.

Recommendation 7.4: NASS and ERS should investigate the feasibility of providing sufficient information on the design and nonresponse characteristics of ARMS, in order to perform design-based statistical analysis without using the replicate weights and to allow users to incorporate design and nonresponse information in model-based analyses.

Recommendation 7.5: ERS should build an enhanced level of in-house survey statistics expertise, in cooperation with NASS. The specialized expertise in both econometrics and survey statistics needed to accomplish this is currently not present in ERS and is likely to require a significant effort in recruiting and training.

Recommendation 7.6: ERS and NASS should collaborate on writing a Guide for Researchers for performing multivariable analyses using data from complex surveys, particularly data from ARMS. In areas in which expertise is not available for writing parts of such a guide, expertise should be sought from the statistics and economics community, especially those with experience in the analysis of survey data from complex survey designs.

Dissemination

Recommendation 8.1: ERS should continue to improve the ARMS web tool by providing summaries on more variables and more subsets from ARMS, and to improve the ARMS extranet web tool by adding the ability to link over years and to more sophisticated models.

Recommendation 8.2: USDA should consider extending the availability of ARMS microdata through the Census Bureau research data centers to increase access opportunities for using additional data sets and enabling researchers to match ARMS files with other data sets.

Recommendation 8.3: ERS should provide more training for new data users, including developing a data user manual, which also includes the recommended guide on statistical estimation, and offering training workshops.

Recommendation 8.4: Database management practices should include a system for managing and reporting errors found by users, for ensuring the consistent labeling of the codes for raw variables, and for using consistent names of the ERS-created summary variables over time.

References

American Association for Public Opinion Research
2006 *Standard definitions: Final dispositions of case codes and outcome rates for surveys.* Lenexa, KS: Author. Available: http://aapor.org/standards.asp.

Baker, R.P., N.N. Bradburn, and R.A. Johnson
1995 Computer-assisted personal interviewing: An experimental evaluation of data quality and cost. *Journal of Official Statistics, 11*, 413-431.

Biemer, P., and G. Forsman
1992 On the quality of reinterview data with application to the Current Population Survey. *Journal of the American Statistical Association, 87*(420), 915.

Binder, D., M. Kovacevic, and G. Roberts
2004 Design-based methods for survey data: Alternative uses of estimating functions. In *ASA Proceedings of the Joint Statistical Meetings* (pp. 3301-3312). Alexandria, VA: American Statistical Association.

Bosecker, R., and W. Kelly
1975 *Summary of results from Nebraska Survey Concept Study.* Washington, DC: National Agricultural Statistics Service.

Bradburn, N.M., S. Sudman, E. Blair, and C. Stocking
1989 Question threat and response bias. In E. Singer and S. Presser (eds.), *Survey Research Methods: A Reader* (pp. 371-384). Chicago: University of Chicago Press.

Breckling, J.U., R.L. Chambers, A.H. Dorfman, S.M. Tam, and A.H. Welsh
1994 Maximum likelihood inference from sample survey data. *International Statistical Review, 62*, 349-363.

Carrington, W., J. Eltinge, and K. McCue
2000 *An economist's primer on survey samples.* Discussion Paper CES-WP-00-15. Center for Economic Studies, U.S. Census Bureau, Washington, DC.

Chambers, R.L.
1986 Design-adjusted parameter estimation. *Journal of the Royal Statistical Society, Series A: General 149*, 161-173.

Charlton, J.
2007 *Towards effective statistical editing and imputation strategies—Findings of the Euredit Project.* Chapter 1, Editing and imputation. Available: http://www.cs.york.ac.uk/euredit/

Claassen, R.
2006 *Conservation policy analysis with ARMS: Challenges and opportunities.* September 28.

Conrad, F.G., and M.F. Schober
2000 Clarifying question meaning in a household telephone survey. *Public Opinion Quarterly, 64,* 1-28.

Conrad, F.G., M.F. Schober, and W. Dijkstra
(in press) Cues of communication difficulty in telephone interviews. In J.M. Lepkowski, C. Tucker, M. Brick, E. de Leeuw, L. Japec, P. Lavrakas, M. Link, and R. Sangster (eds.), *Advances in telephone survey methodology.* New York: Wiley.

Council on Food, Agricultural and Resource Economics
2007 *Improving information about America's farms and ranches: A review of the Census of Agriculture.* Washington, DC: Author.

Covey, T., R. Green, C. Jones, J. Johnson, M. Morehart, R. Williams, C. McGath, A. Mishra, and R. Strickland
2005 *Agricultural income and finance outlook.* Economic Research Service, AIS-83, November.

Culver, D., M. Tomiak, and R.D. Bollman
1991 High and low margin prairie grain and oilseed farms. *Canadian Farm Economics, 23*(1), 49-55.

Curtin, R., S. Presser, and E. Singer
2000 Effects of response rate changes on the index of consumer sentiment. *Public Opinion Quarterly, 64,* 413-428.

Dubman, R.W.
2000 *Variance estimation with USDA's farm costs and returns surveys and Agricultural Resource Management Study surveys.* Staff paper no. AGES 00-01. Resource Economics Division, Economic Research Service, U.S. Department of Agriculture.

DuMouchel, W.H., and G.J. Duncan
1983 Using sample survey weights in multiple regression analyses of stratified samples. *Journal of the American Statistical Association, 78,* 535-543.

Economic Research Service
2003 *The Economic Research Service's information quality guidelines.* Washington, DC: U.S. Department of Agriculture. Available: http://www.ers.usda.gov/AboutERS/QualityGuidelines3.pdf.
2005 *Structure and finances of U.S. farms: 2005 family farm report.* EIB-12. Washington, DC: U.S. Department of Agriculture.
2006a *Structure and finances of U.S. farms: 2005 family farm report.* May. Washington, DC: U.S. Department of Agriculture.
2006b *Water and wetland resources.* AREI, 2006 edition, July. Washington, DC: U.S. Department of Agriculture.
2007 *Farm business and household survey data: Customized data summaries from ARMS.* Washington, DC: U.S. Department of Agriculture. Available: http://www.ers.usda.gov/Data/ARMS/GlobalAbout.htm.

Eklund, B.
1991 *Mobile computerized data entry and edit for the livestock prices received survey.* Research Report No. SRB-91-10, National Agricultural Statistics Service, November. Washington, DC: U.S. Department of Agriculture.
1993 *Computer assisted personal interviews: Experience and strategy.* STB Research Report Number STB-93-01, February. National Agricultural Statistical Service. Washington, DC: U.S. Department of Agriculture.

Esposito, J.L.
2003 *A framework relating questionnaire design and evaluation processes to sources of measurement error.* Federal Committee on Statistical Methodology. Available: http://www.fcsm.gov/03papers/Esposito2.pdf

Federal Committee on Statistical Methodology
2001 *Measuring and reporting sources of error in surveys.* Statistical policy working paper, July 31.

Fowler, F.J., and T.W. Mangione
1990 *Standardized survey interviewing: Minimizing interviewer related error.* Newbury Park, CA: Sage Publications.

Fuller, W.A.
1984 Least squares and related analyses for complex survey designs. *Survey Methodology, 10,* 97-118.

Gerling, M.W.
2005 *Using handheld global positioning system receivers for Phase II of the Agricultural Resource Management Survey.* National Agricultural Statistics Service, October. Washington, DC: U.S. Department of Agriculture.

Golan, E., B. Krissoff, F. Kuchler, L. Calvin, K. Nelson, and G. Price
2004 *Traceability in the U.S. food supply: Economic theory and industry studies.* Agricultural Economic Report No. 830. Economic Research Service, March 2004. Washington, DC: U.S. Department of Agriculture, Available: http://www.ers.usda.gov/Publications/AER830/AER830fm.pdf.

Groves, R.
1987 Research on survey data quality. *Public Opinion Quarterly, 51,* S156-S172.
2006 Nonresponse rates and nonresponse error in household surveys. *Public Opinion Quarterly, 70*(5), 637-645.

Groves, R., S. Presser, and S. Dipko
2004 The role of topic interest in survey participation decisions. *Public Opinion Quarterly, 68,* 2-31.

Hansen, L.
2006 Wetlands: Status and trends. In *Agricultural resources and environmental indicators, 2006 edition* (Chapter 2.3). Economic Research Service, July. Available: http://www.ers.usda.gov/Publications/AREI/EIB16/

Hanuschak, G.A., D. Atkinson, W.C. Iwig, and V. Tolomeo
1991 *History of reinterview studies at NASS.* Joint Statistical Meetings, August 22.

Heckman, J.J.
1979 Sample selection bias as a specification error. *Econometrica, 47*(January), 153-161.

Hellerstein, D., C. Nickerson, J. Cooper, P. Feather, D. Gadsby, D. Mullarkey, A. Tegene, and C. Barnard
2002 *Farmland protection: The role of public preferences for rural amenities.* Agricultural Economic Report No. 815. Economic Research Service. Available: http://www.ers.usda.gov/publications/aer815/aer815.pdf.

Hidiroglou, M., D. Drew, and G. Gray
1993 A framework for measuring and reducing nonresponse in surveys. *Survey Methodology, 19*, 81-94.

Holt, D., T.M.F. Smith, and P.D. Winter
1980 Regression analysis of data from complex surveys. *Journal of the Royal Statistical Society, Series A: General 143*, 474-487.

Hopkins, J., and R. Johansen
2004 Beyond environmental compliance: Stewardship as good business. *Amber Waves*, April.

Horowitz, K., and M. Planting
2006 *Concepts and methods of the U.S. input-output accounts*. Bureau of Economic Analysis. Washington, DC: U.S. Department of Commerce. Available: http://bea.gov/bea/papers/IOmanual_092906.pdf

Juster, F.T., and F.P. Stafford
1991 The allocation of time: Empirical findings, behavioral models and problems of measurement. *Journal of Economic Literature, 29*(June), 471-522.

Juster, F.T., and J.P. Smith
1997 Improving the quality of economic data: Lessons from HRS and AHEAD. *Journal of the American Statistical Association*, 92, 1268-1278.

Kalton, G.
1983 Models in the practice of survey sampling. *International Statistical Review, 51*, 175-188.

Keeter, S., C. Miller, A. Kohut, R. Groves, and S. Presser
2000 Consequences of reducing nonresponse in a large national telephone survey. *Public Opinion Quarterly, 64*, 125-148.

Kennickell, A.B.
1997 *Using range techniques with CAPI in the 1995 Survey of Consumer Finances*. Available: http://www.federalreserve.gov/pubs/oss/oss2/papers/rangepap0197pdf.

Korn, E.L., and B.I. Graubard
1999 *Analysis of health surveys*. New York: John Wiley.

Kott, P.
2003 A plan for coordinating surveys at the National Agricultural Statistics Service. In *Proceedings of Statistics Canada Symposium 2003*. Catalogue no. 11-522-XIE. Ottawa: Statistics Canada.

Kott, P., and M. Fetter
1999 Using multiphase sampling to limit response burden across agricultural surveys. In *Proceedings of the Survey Methods Section*. Ottawa: Statistical Society of Canada.

Lewandrowski, J., M. Peters, C. Jones, R. House, M. Sperow, M. Eve, and K. Paustian
2004 *Economics of sequestering carbon in the U.S. agricultural sector*. Economic Research Service Technical Bulletin No. 1909. Washington, DC: U.S. Department of Agriculture.

Little, R.J.A.
2004 To model or not to model? Competing modes of inference for finite population sampling. *Journal of the American Statistical Association, 99*(466), 546-556.

Little, R.J.A., and D.B. Rubin
2002 *Statistical analysis with missing data*, 2nd edition. Hoboken, NJ: Wiley.

Lohr, S.L.
1999 *Sampling: Design and analysis*. Pacific Grove, CA: Duxbury Press.

Merkle, D., and M. Edelman
2002 Nonresponse in exit polls: A comprehensive analysis. In R. Groves, D. Dillman, J. Eltinge, and R.A. Little (eds.), *Survey nonresponse* (pp. 243-257). New York: Wiley.

Moore, J., L. Stinson, and E. Welniak
1997 Income measurement error in surveys: A review. In M. Sirken, D. Herrmann, S. Schechter, N. Schwarz, J. Tanur, and R. Tourangeau (eds.), *Cognition and survey research* (pp. 155-174). New York: Wiley.

National Agricultural Statistics Service
2005 *Supporting statement, agriculture resource management, chemical use, and post-harvest chemical use surveys*. OMB No. 0535-0218.
2007a *Information quality guidelines*. Available: http://www.usda.gov/nass/nassinfo/infoguide.htm.
2007b *Education and outreach: Research reports*. Available: http://www.nass.usda.gov/Education_&_Outreach/Reseach_Reports/index.asp.
2007c *Agricultural Resource Management Survey overview*. Available: http://www.usda.gov/nass/nassinfo/surveyprograms/tables1.htm#agresource.
2007d *Data and statistics: Special tabulations*. Available: http://www.nass.usda.gov/Data_and_Statistics/Special_Tabulations/index.asp.

National Association of State Departments of Agriculture
2007 *NASDA organizational profile*. Available: http://www2.nasda.org/NR/rdonlyres/DB09E55B-3101-475F-A665-423538F9EE83/992/NASDAOrganizationalProfile-Feb2007.pdf.

Nathan, G., and T.M.F. Smith
1989 The effect of selection on regression analysis. In C.J. Skinner, D. Holt, and T.M.F. Smith (eds.), *Analysis of complex surveys* (pp. 149-163). New York: Wiley.

Neas, K., J. Molina, J. Hardegree, and M.W. Gerling
2006 *Using personal digital assistants for the 2004 Cotton Objective Yield Survey.* National Agricultural Statistics Service, March.

Nordberg, L.
1989 Generalized linear modeling of sample survey data. *Journal of Official Statistics, 5*, 223-239.

Organisation for Economic Co-operation and Development
2003 *Quality framework and guidelines for OECD statistical activities*. Version 2003/1, October 17.

Oh, H.L., and F.J. Scheuren
1983 Weighting adjustments for unit non-response. In W.G. Madow, I. Olkin, and D.B. Rubin (eds.), *Incomplete data in sample surveys*, Volume 2 (pp. 143-184). New York: Academic Press.

Ott, K.
1999 *Panel Plus Pilot Survey: Results from the 1998-1999 ARMS Panel Plus Pilot Survey*. Research Division Research Report No. RD-99-05, December.

Ott, K., and C. Crouse
2005 Professionally designed questionnaires: Do they collect better data? *Proceedings of the American Statistical Association Section on Research Survey Methods*, Alexandria, VA: American Statistical Association.

Perry, C.R., J.C. Burt, and W.C. Iwig
1994 Redrawing the 1993 Farm Costs and Returns Survey list sample to reduce its overlap with three other 1993 surveys and the 1992 FCRS. *ASA Proceedings of the Section on Survey Research Methods*, 632-637.

Pfeffermann, D.
1993 The role of sampling weights when modeling survey data. *International Statistical Review, 61*, 317-337.

Pfeffermann, D., and M. Sverchkov
1999 Parametric and semi-parametric estimation of regression models fitted to survey data. *Sankhya,* B61:166-186.

Polivka, A.E., and J.M. Rothgeb
1993 Redesigning the CPS questionnaire. *Monthly Labor Review,* 10-28.

Press, S., and J. Tanur
2000 Respondent-generated interval estimation to reduce item-nonresponse. In *Applied Statistical Science V,* 39-48. Huntington, NY: Nova Science Publishers.

Rao, J.N.K.
2005 Interplay between sample survey theory and practice: An appraisal. *Survey Methodology, 31,* 117-138.

Rao, J.N.K., C.F.J. Wu, and K. Yue
1992 Some recent work on resampling methods for complex surveys. *Survey Methodology, 18,* 209-217.

Ribaudo, M., and R. Johansson
2006 Water quality: Impacts on agriculture. In *Agricultural resources and environmental indicators,* 2006 edition (Chapter 2.2). Economic Research Service, July. Available: http://www.ers.usda.gov/Publications/AREI/EIB16/.

Rubin, D.B.
1976 Inference and missing data. *Biometrika, 63,* 581-590.

Rust, K.F., and J.N.K. Rao
1996 Variance estimation for complex surveys using replication techniques. *Statistical Methods in Medical Research, 5,* 283-310.

Sarndal, C.E., and S. Lundstrom
2005 *Estimation in surveys with nonresponse.* Chichester, UK: Wiley.

Sarndal, C.E., and B. Swensson
1987 A general view of estimation for two phases of selection with applications to two-phase sampling and nonresponse. *International Statistical Review, 55,* 279-294.

Sarndal, C.E., B. Swensson, and J. Wretman
1992 *Model assisted survey sampling.* New York: Springer-Verlag.

Schober, M.F., and Conrad, F.G.
1997 Does conversational interviewing reduce survey measurement error? *Public Opinion Quarterly, 61,* 576-602.

Schober, M.F., F.G. Conrad, and S.S. Fricker
2004 Misunderstanding standardized language in research interviews. *Applied Cognitive Psychology, 18,* 169-188.

Skinner, C.J.
1994 Sample models and weights. In *ASA Proceedings of the Section on Survey Research Methods* (pp. 133-142). Alexandria, VA: American Statistical Association.

Skinner, C.J., D. Holt, and T.M.F. Smith
1989 *Analysis of complex surveys.* New York: Wiley.

Stanley, J.
1996 *Standardizing interviewer behavior based on the results of behavior coding.* Paper presented at the 51st annual conference of the American Association of Public Opinion Research, Salt Lake City.

Statistics Canada
2003 *Statistics Canada quality guidelines,* fourth edition. Catalogue No. 12-539-XIE, October. Ottawa: Author.

Tourangeau, R., and T.W. Smith
1996 Asking sensitive questions: The impact of data collection mode, question format, and question context. *Public Opinion Quarterly* 60:275-304.

U.S. Census Bureau

2003 *Census Bureau standard: Pretesting questionnaires and related materials for surveys and censuses*. July 25.

U.S. Code

USC Title 7, Chapter 35A, II, 1441a.

U.S. Department of Agriculture

1944 *Income parity for agriculture, part 1: Farm income, bureau of economic analysis*. Washington, DC: Author.

2006a *Making a difference for America's small farmers and ranchers in the 21st century*. Small Farms Coordinators, January. Washington, DC: Author.

2006b *USDA quality of information guidelines*, June 28. Available: http://www.ocio.usda.gov/qi_guide/index.html.

U.S. Office of Management and Budget

2001 *Guidelines for ensuring and maximizing the quality, objectivity, utility, and integrity of information disseminated by federal agencies*. September 28.

2002 Federal statistical organizations' guidelines for ensuring and maximizing the quality, objectivity, utility, and integrity of disseminated information. *Federal Register*, 67(107).

2006a *Standards and guidelines for statistical surveys*. September.

2006b *Standards and guidelines for statistical surveys*. October.

Van Schaik, C.

2003 *Getting a handle on the barriers to financing sustainable agriculture: The gaps between farmers and lenders in Minnesota and Wisconsin*. Land Stewardship Project. Available: http://www.landstewardshipproject.org.

Wolter, K.M.

1985 *Introduction to variance estimation*. New York: Springer-Verlag.

Yung, W., and J.N.K. Rao

1996 Jackknife linearization variance estimators under stratified multi-stage sampling. *Survey Methodology*, 22, 23-31.

Zieschang, K.D.

1990 Sample weighting methods and estimation of totals in the consumer expenditure survey. *Journal of the American Statistical Association*, *85*, 986-1001.

Appendix A

Panel Data-Gathering Activities

The panel conducted the bulk of its business in open forum, eliciting input from staff of the Economic Research Service (ERS) and the National Agricultural Statistics Service (NASS) as well as key data users, policy makers, and additional technical experts. The panel's public sessions included a workshop on statistical methodology, June 8-9, 2006; a session for data users at the annual meeting of the American Agricultural Economics Association, July 24, 2006; a workshop on concepts and measurement, September 28-29, 2006; a workshop on inference, December 7-8, 2006; and an open discussion of cost-of-production issues, January 18, 2007.

At the panel's introductory meeting in February 2006, Katherine Smith, then associate director of ERS, and Carol House, associate director of NASS, set the stage for the work of the panel. They discussed the purpose of the study from their perspective, and the charge to the panel. Their frank and open exposition on the need for a review of the survey were instrumental in focusing our attention on key issues to be addressed in the study and in enhancing our understanding of the partnership between these agencies that directs and sustains the survey. In Carol House's presentation, the panel was challenged to consider not only the question of whether the agencies were doing the right things in terms of content, but also whether they are doing things correctly in terms of approach and methodology. Katherine Smith emphasized the critical importance of the survey as the main source of information on farm economics, production, and structure.

These presentations were supported and elaborated by a thorough discussion of the state of the survey by Robert Bass and Mary Bohman, who addressed issues arising from the complexity of the Agricultural Resource Management Survey (ARMS): the complicated survey form, the problem of missing observations and nonresponse, the lack of comparability of measures across subsamples, and the difficulties in analysis imposed by the complex estimation and other procedures, among other topics.

At our second meeting, the panel focused on issues of statistical methodology and heard mainly from staff of NASS and ERS with responsibility for various aspects of survey design and data collection, estimation, and processing. In the discussion of questionnaire content, the panel learned from Doug Kleweno, Jim Johnson, Carol Jones, and Bill McBride about the determinants of the content of ARMS from the perspective of resource use, farm business performance, farm household well-being, and commodities, respectively. Two presentations were made on the process of turning concepts into questions, as well as the methods of testing and implementing the questionnaires by Danna Moore of the University of Washington, and Kathy Ott of NASS. Collection methodology and other statistical aspects of the survey were discussed by Bob Bass, Bill Iwig, Alix Riley, Richard Barton, and Chadd Crouse. In the final session, Phillip Kott, Bob Dubman, and Nigel Key introduced the challenging topic of methods of imputation and analysis of the ARMS data. The panel revisited the difficult issues involved in imputation with a complex survey design in our fourth meeting.

At the annual meeting of the American Agricultural Economics Association (AAEA) in August 2006, several panel members met informally with a large group of frequent ARMS data users to discuss accessibility of the data and data dissemination. The users also introduced issues regarding response rates, survey design, content, classification, editing and imputation, coding, cross-survey comparisons, and ongoing user forums. A summary of the discussion was reported to the entire panel at our third meeting.

The third meeting focused on the needs of major public- and private-sector data users and on data dissemination. Useful insights on data needs by major federal users were provided by Jim Langley, Joseph Cooper, and Joseph Glauber, representing the perspective of the Congressional Budget Office, the Council of Economic Advisers, and the Office of the Chief Economist, U.S. Department of Agriculture (USDA), respectively. Bob Young discussed the data needs of the general interest U.S. American Farm Bureau Federation, and Gary Adams laid out the requirements of the statistics program of the National Cotton Council, one of the major commodity-oriented interest groups. Discussing concepts and measures for understanding environmental analysis, commodities, and farm household well-being were Timothy Kiely and Arthur Grube, U.S. Environmental Protection Administration; Jeff Goebel, National Resources Conservation Service, USDA; and Roger Claasen, Bill McBride, Carol Jones, and Jim Johnson, ERS.

The panel conducted a lively discussion of issues involved in dissemination of survey data, particularly to users of individual record data for research purposes, supported by presentations by Mitch Morehart, ERS, and Ron Jarmin, U.S. Census Bureau.

Two substantive topics were revisited in the fourth and fifth panel meetings, which otherwise were largely focused on report preparation and discussion of potential recommendations. The panel delved more deeply into the problem of data analysis for complex surveys, with presentations by William Greene, New York University, who served as a consultant to the panel, Phillip Kott, and Jim MacDonald, ERS, who elaborated earlier presentations.

Bill McBride rejoined the panel to add detail to prior discussions of the treatment of commodities in ARMS, with a focus on the implementation of recommendations by a special AAEA panel on cost of production.

Agendas of the panel's data-gathering meetings appear below.

MEETING AGENDAS

First Meeting
February 2-3, 2006

Goals for the first meeting:

Introduce panel members to each other and to the supporting staff.
Conduct the bias and conflict of interest discussion.
Hold an open session with representatives of the sponsoring agency and other experts to discuss the charge to the panel and to learn more about the Agricultural Resource Management Survey.
Organize the approach to the task.
Set dates for future meetings.

Thursday, February 2

12:00 – 1:00 p.m.	**Welcome and Panel Introductions** Bruce Gardner, Chair Michael Feuer, Executive Director, Division of Behavioral and Social Sciences and Education Constance Citro, Director, Committee on National Statistics

1:00 – 1:45 p.m.	**Overview of the Agenda** Bruce Gardner, Chair **Begin Discussion of the Statement of Work** Panel Discussion
1:45 – 2:00 p.m.	**Break**
2:00 – 3:00 p.m.	**Charge to the Panel** Kitty Smith, Associate Director, Economic Research Service Carol House, Associate Director, National Agricultural Research Service
3:00 – 4:30 p.m.	**Status of the Agricultural Resource Management Survey/Issues to Be Addressed** Mary Bohman, Director, Resource Economics Division, ERS Robert Bass, Director, Census and Survey Division, NASS
4:30 – 5:00 p.m.	**Bias and Conflict of Interest Discussion** Kirsten Sampson-Snyder
5:00 – 5:45 p.m.	**Discussion of ARMS Program Issues** Bruce Gardner, Chair
5:45 p.m.	**Adjourn for the Day**

Friday, February 3

8:30 – 10:00 a.m.	**Working Session** • Review information sharing exercise from prior day; identify needed adjustments to approach • Discuss issues to be examined; develop plans to address each; assign panel responsibilities • Develop plans for follow-on activities

10:00 – 10:15 a.m.	**Break**
10:15 a.m. – 12:00 p.m.	**Working Session Continues**
12:00 – 1:00 p.m.	**Lunch**
1:00 – 3:00 p.m.	**Working Session** • Revisit adequacy of panel expertise • Determine if consultant assistance is needed • Discuss audience for final report • Develop outline of final report • Develop plans for future meetings
3:00 p.m.	**Adjourn**

SECOND MEETING

June 8-9, 2006

Goals for the statistical methodology workshop:
To discuss statistical methodology issues in the ARMS program.
To plan for the upcoming Concepts and Measurements workshop.
To develop an outline for the final report.

Thursday, June 8

8:30 – 9:00 a.m.	**Panel Business** Bruce Gardner, Chair
	Bias and Conflict of Interest Discussion Kirsten Sampson-Snyder
9:00 – 9:15 a.m.	**Welcome and Overview of the Agenda** Bruce Gardner, Chair
9:15 – 10:00 a.m.	**Interagency Interaction on Issues of Statistical Methodology**

	Robert Bass, Director, Census and Survey Division, NASS
	Mary Bohman, Director, Resource Economics Division, ERS
10:00 – 10:15 a.m.	**Break**
10:15 a.m. – 12:00 p.m.	**Survey Management and Data Collection Issues and Statistical Methodology Issues: Questionnaire Content** *Cynthia Clark, Moderator* Information on Resource Use and Allocation at the Field Level Doug Kleweno, NASS Information on Farm Business Performance: Resources, Environment, Other Enterprise Information Jim Johnson, ERS Information on the Farm Household: Economic Well-Being Carol Jones, ERS Information on Commodities Bill McBride, ERS
12:00 – 1:00 p.m.	**Lunch**
1:00 – 5:30 p.m.	**Survey Management and Data Collection Issues and Statistical Methodology Issues (Continued)** *Fred Conrad, Moderator* Questionnaire Design TBA Questionnaire Development Kathy Ott, NASS Collection Methodology—Current and Options (CATI, CAPI, web-based); Recruiting and Training of Interviewers Bob Bass, NASS
5:30 p.m.	**Adjourn for the Day**

Friday, June 9

8:30 – 10:00 a.m.	**Survey Management and Data Collection Issues and Statistical Methodology Issues (Continued)** *Arthur Kennickell, Moderator* Sample Design and Frame (list and area) Bill Iwig, NASS Nonresponse (unit and item) Alix Riley, NASS Estimation and Imputation Richard Barton, NASS Nonresponse and Calibration Chadd Crouse, NASS
10:00 – 10:15 a.m.	**Break**
10:15 a.m. – 12:00 p.m.	**Survey Management and Data Collection Issues and Statistical Methodology Issues: Methods for Analysis** *David Binder, Moderator* Complex Design; Jackknife Phil Kott, NASS ERS Procedures for Statistical Analysis Bob Dubman, ERS Scientific Uses in Research Nigel Key, ERS Review of Statistical Packages for Improving Analysis TBA
12:00 – 1:00 p.m.	**Lunch**
1:00 – 3:00 p.m.	**Working Session** • Review information gathered during workshop • Develop outline of final report • Develop plans for upcoming concepts and methods workshop
3:00 p.m.	**Adjourn**

THIRD MEETING

September 28-29, 2006

Goals for the concepts and measurements workshop:

- To discuss conceptual and measurement issues in the ARMS program.
- To explore the needs of data users.
- To consider possible improvements in data dissemination, particularly for protected dissemination of sensitive individual records for research purposes.
- To explore issues regarding tests of inference for regression analysis using ARMS data.
- To further refine the outline for the final report and make plans for the upcoming meeting on December 7-8, 2006.

Thursday, September 28

8:30 – 9:00 a.m.	**Panel Business** Bruce Gardner, Chair
9:00 – 9:15 a.m.	**Welcome and Overview of the Agenda** Bruce Gardner, Chair
9:15 – 10:00 a.m.	**Report on the AAEA Data Users Forum** Bruce Gardner, Panel Members
10:00 – 10:15 a.m.	**Break**
10:15 a.m. – 12:00 p.m.	**Data Uses and Users**
12:00 – 1:00 p.m.	**Lunch**
1:00 – 5:30 p.m.	**Concepts and Measures**
5:30 p.m.	**Adjourn for the Day**

Friday, September 29

8:30 – 10:00 a.m.	**Tests of Inference for Regression Analysis**
10:00 – 10:15 a.m.	**Break**

10:15 a.m. – 12:00 p.m.	**Discussion of Methods of Analysis Issues** *David Binder, Moderator*
12:00 – 1:00 p.m.	**Lunch**
1:00 – 3:00 p.m.	**Working Session** • Review information gathered during workshop • Develop outline of final report • Develop plans for upcoming December 7-8 meeting
3:00 p.m.	**Adjourn**

FOURTH MEETING

December 7-8, 2006

Goals for the inference workshop:

To discuss issues regarding tests of inference for data analysis using ARMS data.
To further refine the outline for the final report.
To discuss preliminary findings and recommendations.
To make assignments for tasks for completing the final report.
To make plans for the upcoming meeting on January 18-19, 2007.

Thursday, December 7

8:30 – 8:45 a.m.	**Introduction to Problem of Inference and Summary of Prior Discussion on the Topic** Bruce Gardner, Chair
8:45 – 9:30 a.m.	**Issues in Data Analysis for Complex Surveys** William Greene, New York University (Consultant)
9:30 – 10:00 a.m.	**Alternatives to Current Methods of Estimation and Variance Computation** David Binder, Discussion Leader

10:00 – 10:15 a.m.	**Break**
10:15 – 10:45 a.m.	**Practical Issues in Applying Methods of Inference to Complex Data Sets** Bill Greene, Discussion Leader
10:45 – 11:00 a.m.	**Summary of Issues in Complex Data Analysis** Bruce Gardner, Chair
11:00 a.m. – 12:00 p.m.	**General Discussion of ARMS Issues** Bruce Gardner, Chair
12:00 – 1:00 p.m.	**Lunch**
1:00 – 5:30 p.m.	**Working Session** Panel discussion of issues
5:30 p.m.	**Adjourn for the Day**

Friday, December 8

8:30 – 10:00 a.m.	**Working Session** Review of information gathered Refine outline for final report
10:00 – 10:15 a.m.	**Break**
10:15 a.m. – 12:00 p.m.	**Working Session** Assignments for preparation of final report
12:00 – 1:00 p.m.	**Lunch**
1:00 – 3:00 p.m.	**Working Session** • Plans for meeting on January 18-19, 2007
3:00 p.m.	**Adjourn**

FIFTH MEETING

January 18-19, 2007

Goals for the meeting:

- To discuss issues regarding cost of production estimates from the ARMS survey with ERS/NASS staff.
- To further refine the final report.
- To further develop sections written since the December meeting.
- To make additional assignments for tasks for completing the final report.
- To validate initial findings and recommendations.
- To make plans for the upcoming meeting in March 2007.

Thursday, January 18

8:30 – 9:00 a.m.	Discussion of Approach to this Meeting Bruce Gardner, Chair
9:00 – 11:00 a.m.	Discussion of AAEA Cost of Production Report and Status of Implementation of Recommendations in ARMS Bill McBride, ERS
11:00 a.m. – 12:00 p.m.	General Discussion of ARMS Issues with Department of Agriculture Staff Bruce Gardner, Chair
12:00 – 1:00 p.m.	Lunch
1:00 – 2:30 p.m.	Break-out Working Sessions
2:30 – 5:30 p.m.	Panel Discussion of Issues
5:30 p.m.	Adjourn for the Day

Friday, January 19

8:30 – 10:00 a.m.	Working Session Refine Final Report

10:00 – 10:15 a.m.	**Break**
10:15 a.m. – 12:00 p.m.	**Working Session:** Assignments for Final Report
12:00 – 1:00 p.m.	**Lunch**
1:00 – 3:00 p.m.	**Working Session** Plans for Meeting in March
3:00 p.m.	**Adjourn**

Appendix B

Biographical Sketches of Panel Members and Staff

BRUCE GARDNER (*Chair*) is professor and interim dean of the College of Agriculture at the University of Maryland. Previously he was a faculty member at Texas A&M University and North Carolina State University. During 1975-1977, he was a senior staff economist with the Council of Economic Advisers, covering agricultural issues during the time of the first Soviet grain trade agreement and the development of the 1977 Farm Bill in the Carter administration. During 1989-1992, he was assistant secretary of agriculture and the chief economist at the U.S. Department of Agriculture. His writings have concentrated on agricultural commodity and trade policy, marketing, and farm income distribution and have received three awards for excellence from the American Agricultural Economics Association. He is a fellow of the association and was its president in 1999. He has a B.S. in agricultural science and economics from the University of Illinois and a Ph.D. in economics from the University of Chicago.

WALTER J. ARMBRUSTER is president of the Farm Foundation in Oak Brook, Illinois. He has previously worked at the U.S. Department of Agriculture as a staff economist involved with research in marketing efficiency, institutions, and policy issues. He is past president and a fellow of the American Agricultural Economics Association and is secretary/treasurer of the International Association of Agricultural Economists. He has served on the advisory board of the National Agricultural Statistics Service. He has B.S. and M.S. degrees in agricultural economics from Purdue University and a Ph.D. in agricultural economics from Oregon State University.

DAVID BINDER is a consultant on survey design and was previously director general of the methodology branch of Statistics Canada. He is a fellow of the American Statistical Association and an elected member of the International Statistical Institute. He has written extensively on the theory and methods for the analysis of complex survey data. At the National Research Council, he served on the Panel on the Research on Future Census Methods. He is a past president of the Statistical Society of Canada. He has a B.Sc. from the University of Toronto and a Ph.D. from Imperial College, London.

RAY D. BOLLMAN is chief of research and analysis of the agriculture division of Statistics Canada. He also serves as adjunct research professor at the University of Manitoba. He has written extensively on rural development issues and has analyzed data sets on household expenditure patterns for rural families. He has B.Sc. and M.Sc. degrees from the University of Manitoba and a Ph.D. in economics from the University of Toronto.

CYNTHIA Z.F. CLARK is an executive director at the U.K. Office for National Statistics leading the Methodology Directorate. Previously she served as associate director for methodology and standards at the U.S. Census Bureau where she led the large scale evaluation of the 2000 census, established an administrative records research program, initiated a framework for quality standards, and developed a usability laboratory. She has 13 years experience with agriculture surveys and censuses as a division director at the National Agricultural Statistics Service and research manager in the Agriculture Division of the Census Bureau. Her professional work focuses on survey and census methodology, operations, and research; official statistics; survey technologies; privacy and confidentiality; and statistical training in the workplace. She is a fellow of the American Statistical Association and an elected member of the International Statistical Institute. She has M.S. and Ph.D. degrees in statistics from Iowa State University.

FREDERICK CONRAD is research associate professor of the Institute for Social Research at the University of Michigan. His research generally involves identifying and reducing survey measurement error by applying ideas and methods from cognitive science. His current research is focused on adaptive user interfaces in web surveys, understanding and misunderstanding survey questions, estimation processes, evaluation of questionnaire pretesting methods, and interviewer-respondent interaction. He has a Ph.D. in psychology from the University of Chicago.

ANI L. KATCHOVA is assistant professor in the Department of Agricultural and Consumer Economics of the University of Illinois at Urbana-

Champaign, where she teaches applied statistical methods, and advanced agricultural economics. Her recent publications have made extensive use of microdata from the Agricultural Resource Management Survey, and her research has focused on agricultural finance with a focus on farm financial performance. She has a Ph.D. in agricultural economics from the Ohio State University.

ARTHUR KENNICKELL is a senior economics and unit head for the Survey of Consumer Finances of the Board of Governors of the Federal Reserve System. He has been on the staff of the Board of Governors of the Federal Reserve System since 1984. His areas of expertise are data collection and estimation methodology, microeconomics, and macroeconomics. He is a fellow of the American Statistical Association. He has B.A. and M.A. degrees from the University of Chicago and a Ph.D. in economics from the University of Pennsylvania.

CATHERINE KLING is professor of economics in the Department of Economics at Iowa State University. She also serves as head of the resource and environmental policy division in Center for Agricultural and Rural Development there. She has been on the staff of Iowa State University since 1993. She serves as associate editor of the Journal of Environmental Economics and Management. Her research has focused on issues of agricultural environmental and resource economics. She has a Ph.D. in economics from the University of Maryland and a B.B.A. in business and economics from the University of Iowa.

CARYN KUEBLER (*Associate Program Officer*) is an associate program officer for the Committee on National Statistics. Previously she worked for the University of Chicago's Cultural Policy Center on a nationally scaled research project measuring the relationship between the size and scope of a region's creative sector and its economic growth potential. Her research interests include measuring consumer debt burden and income inequality, economic development, and cultural policy, including access to and protection of cultural and natural resources. She has a B.S. from Syracuse University and an M.P.P. from the University of Chicago.

JEAN OPSOMER is director of the Center for Survey Statistics and Methodology and associate professor in the Department of Statistics at Iowa State University. His research interests and professional practice include nonparametric regression; developing advanced statistical tools to increase understanding of environmental processes, as well as the human impact on the environment; and the design and estimation for the National Resources Inventory survey. He has an M.S. in management engineering from

Katholieke Universiteit, Leuven, Belgium, an M.B.A. from the University of Chicago, and a Ph.D. in operations research from Cornell University.

BOBBY R. PHILLS is professor and director of the small fruits program at Florida A&M University. He has directed several research projects for the U.S. Department of Agriculture's Agricultural Research Service and serves as a member of the advisory board of the National Agricultural Statistics Service. He is a member of the National Research Council's Board on Agriculture and Natural Resources. He has a B.S. in horticulture from Southern University and A&M College, an M.S. in horticulture from Louisiana State University, and a Ph.D. in horticulture/plant breeding from Louisiana State University.

THOMAS J. PLEWES (*Study Director*) is a senior program officer for the Committee on National Statistics and was study director for its Panel to Review Research and Development Statistics at the National Science Foundation. Previously he was associate commissioner for employment and unemployment statistics of the Bureau of Labor Statistics and served as chief of the U.S. Army Reserve. He was a member of the Federal Committee on Statistical Methodology. He is a fellow of the American Statistical Association. He has a B.A. in economics from Hope College and an M.A. in economics from the George Washington University.

STANLEY PRESSER is professor in the Sociology Department and the Joint Program in Survey Methodology at the University of Maryland. His research has focused on the interface between social psychology and survey measurement, questionnaire design and testing, the accuracy of survey responses, and ethical issues stemming from the use of human subjects. He is a past president of the American Association for Public Opinion Research and is a fellow of the American Statistical Association. At the National Research Council, he served on the Committee to Review the Bureau of Transportation Statistics' Survey Programs. He has a Ph.D. in sociology from the University of Michigan.

ROBERT D. TORTORA is chief methodologist of the Gallup Organization. Prior to joining Gallup, he was associate director for statistical design, methodology, and standards at the U.S. Bureau of the Census. In that position, he led the design of the 2000 census; he was responsible for the statistical and survey methodology for all Census Bureau programs, as well as the development of statistical and survey standards. Prior to joining the Census Bureau staff, he was director of the research and applications division of the National Agricultural Statistics Service. He also serves as adjunct professor of statistics at the University of Nebraska and of applied statistics at George Mason University. He is a fellow of the American Statistical Association. He has a B.S. in mathematics from Youngstown State University, an M.S. in mathematical statistics from the Catholic University of America, and a Ph.D. in probability and statistics from Bowling Green State University.

MICHAEL K. WOHLGENANT is William Neal Reynolds distinguished professor of agricultural and resource economics at North Carolina State University. His specialty is development of economic models of agricultural marketing, policy, and price-analysis problems. He has developed economic models to understand farm-to-retail price linkages, consumer preferences, and the effects of advertising. He has had extensive commodity experience, including work on applied price and marketing problems for cotton, dairy, beef, pork, grapes, sugar, tobacco, wine, and horticultural crops. He has extensive experience in modeling the impact of generic advertising on farm-level returns to producers. He has also made contributions to understanding the allocation of check-off funds between research and advertising. He was an innovator of the equilibrium displacement modeling approach, which is used extensively in policy and welfare analysis. He has B.S. and M.S. degrees in economics and agricultural economics from Montana State University and a Ph.D. in agricultural economics from the University of California at Davis.

COMMITTEE ON NATIONAL STATISTICS

The Committee on National Statistics (CNSTAT) was established in 1972 at the National Academies to improve the statistical methods and information on which public policy decisions are based. The committee carries out studies, workshops, and other activities to foster better measures and fuller understanding of the economy, the environment, public health, crime education, immigration, poverty, welfare, and other public policy issues. It also evaluates ongoing statistical programs and tracks the statistical policy and coordinating activities of the federal government, serving a unique role at the intersection of statistics and public policy. The committee's work is supported by a consortium of federal agencies through a National Science Foundation grant.